T5-ARZ-415

PRECIOUS METALS
A Guide to Trading and Investing

PRECIOUS METALS

A Guide to Trading and Investing

by Anthony George Gero

Lyle Stuart Inc. *Secaucus, New Jersey*

Published by Lyle Stuart Inc.
Published simultaneously in Canada by
Musson Book Company,
A division of General Publishing Co. Limited
Don Mills, Ontario

Queries regarding rights and permissions should be
addressed to: Lyle Stuart, 120 Enterprise Avenue,
Secaucus, N.J. 07094

Manufactured in the United States of America

1 2 3 4 5

Library of Congress Cataloging in Publication Data

Gero, Anthony George.
Precious metals.

Bibliography: p.
Includes index.
1. Precious metals. 2. Metals as an investment.
I. Title.
HG261.G47 1985 332.63 85-2631
ISBN 0-8184-0371-3

Acknowledgments

I would like to thank my fellow traders and members of the exchanges, especially Alan J. Brody, president of the Commodity Exchange (COMEX); and Oscar Burchard, chairman of COMEX; Michel Marks, chairman of the New York Mercantile Exchange (NYMEX); NYMEX's Rosemary McFadden, the most attractive president of any commodity exchange; and Charles S. Horgan, general counsel of NYMEX. Special thanks to Jean LeBreton, secretary of NYMEX; Paul Sarnoff, noted authority on options; and William F. X. Scheinman, author of *Why Most Investors Are Mostly Wrong Most of the Time*. A debt of gratitude as well to friends from the NYMEX Board of Governors and fellow committee members of COMEX and NYMEX, notably Fred Horn, Dr. Henry Jarecki, Moses Marx, Irving Redel, Shep Shaff, Dennis Suskind and Alex Weissenborn. At the COMEX, staffers Julie Gross, Martin Mosbacher and Matt Zachowski have always been of service to me, as have Dave Lieberman and the rest of the compliance staffers who helped me review the rules and regulations of both NYMEX and COMEX. Carlos Rifon, from the Commodity Futures Trading Commission (CFTC). My assistant, Don Allen, and my secretary Dawn Martin, deserve my salute. Jed Horowitz and Kathy Anday edited tirelessly. Claire Beneson, from the New School for Social Research, has been my own special friend in this effort. And I cannot forget my own teachers, the brokers of NYMEX and COMEX who ten years ago patiently showed me the ropes: the Augellos, the Buccellato brothers, the Di Liberto brothers, the Edelstein brothers, Johnny Ezzo, Joe and Bucky Friedman, John Hannemman, Pepper Jones, the Kaufman brothers, the Kaplans, Mel Lazarus, the Lisi and Rowland brothers, Al London, Jimmy Mc Hale, the Mayers, the Meierfelds, John Morace, Pete Morando, Owen Morrissey, the Massa brothers, Nestor Pereira, Frank Polanish, Richie Pront, Willie Rodriguez, the Saitta brothers, H.A. Schwartz, the Tucks, Leo Walsh, Stevie Willner and Carmine, Ira Kishlik, Dominick, Tony and Vinni, Gunther and Lance.

I would also like to thank Eddie Karmin, Mike Belmont and Robert Saitta, three who left us with a void and whom we miss dearly.

Contents

Preface

Gold and precious metals can be owned in several different ways. It is possible to buy shares of a mutual fund that purchases precious metals or that invests in precious metals corporations of one or many nations, such as South Africa or Canada. One can also invest in precious metals options contracts—where you pay for the *right* but not the obligation to buy or sell—coins (minted by many nations and including the popular South African Krugerrand and the Canadian maple leaf), bullion certificates, and bullion and numismatic coins.

Gold mutual funds are perhaps the most common investments. A variety of dealers and brokers offer their own gold funds. Very few funds, though, are attractive investments because in purchasing a gold fund a number of charges are incurred. These include commissions, storage costs and insurance. Funds vary in their strategy from strict investment in gold bullion to investments in a variety of precious metal vehicles.

Most funds tend to concentrate heavily on mining shares. When investing in mining shares, you are investing in the political uncertainties, currency uncertainties, regulatory uncertainties and, certainly, the labor uncertainties in gold mining countries. Therefore, unless gold mining shares have a much higher than available yield compared to other markets, mining shares are not the best possible investment at most times.

There are times, however, when mining shares become very attractive. But beware of whom you buy from in those times, because many people have been burned in fly-by-night operations that open shop when gold seems to glitter most brightly.

The pitfalls of owning mining shares should lead most investors to restrict themselves to investing only through reputable brokerage firms that specialize in precious metals research. Bear in mind that one major advantage of a gold fund is that you can participate with a relatively small amount of money. I think that prior to investing in any sort of a gold fund, one must not only read carefully the prospectus that describes the operation and portfolio of the fund, but should talk to various research departments of brokerage firms. A full-service broker who can provide you with research and guidance is usually preferable to one who gives few of these services while touting its low commissions or high-leverage opportunities.

Another precious metal investment is jewelry. I will give this short shrift because of jewelry's negative investment aspects. Jewelry as an investment suffers from high mark-up, low liquidity, difficult negotiability, indirect investment and high taxation. Jewelry also is highly visible, making it vulnerable to lose either by theft or misplacement. So much for jewelry.

How much should you invest in precious metals? This is a very difficult decision. It should be made on the basis that you never invest more than you can afford to lose.

A recommended portfolio for most clients is one divided three ways, much like the Mercedes-Benz symbol. One-third of the portfolio should be in high-grade interest-bearing obligations providing a very high return of capital (such as highly rated bonds). Another can be in high-grade securities of all sorts, depending on one's outlook on the securities markets. The final third should be in some sort of precious metals investment.

The reason for the triangular portfolio is to diversify the risk. If you have a portfolio recommended by a recognized dealer who follows this strategy, you will usually change the ratios only when one portfolio sector falls below certain recommended standards.

If you have determined that you want to maintain one-third of your investment in precious metals and you find the ratio rises above that due to price appreciation or the additional acquisition of precious metals through fund ownership, a reduction in that investment is warranted. On the other hand, if an increase occurs in another part of the portfolio, more precious metals could be purchased. A professional money manager or broker will help you maintain the right balance.

In the precious metals field, one can continually adopt a stance that is either aggressive—by holding outright positions in the metals—or defensive—by hedging the precious metals portfolio through the sale of other precious metals. It is important to maintain the aggressive-defensive stance at a predetermined percentage.

The one precious metals purchase method I have left out thus far is the futures, or commodity, markets. It is my favorite forum, and is the marketplace where I practice my profession. This book will give you insight into the complex but exciting world of the precious metals commodities market.

PRECIOUS METALS
A Guide to Trading and Investing

ONE

My Life as a Trader

I trade precious metals for a living at the Commodity Exchange Center (CEC) at 4 World Trade Center in New York City. Every morning at 8:45 A.M. I take up my station on the bottom step of the southeast center of the gold ring on the COMEX (the Commodity Exchange, hereinafter written Comex) quadrant of the CEC. (The CEC houses Comex, the New York Mercantile Exchange, the New York Cotton Exchange and the Coffee, Sugar and Cocoa Exchange. More than 85 percent of all metals futures in the United States are traded on Comex.) I trade on both Comex and NYMEX. My position affords me a view of the NYMEX platinum board and the Comex silver and copper boards to my immediate right. These flash up-to-the-minute prices, and daily highs and lows. News and price quote monitors are immediately in front of me. To my left stand brokers from the major precious metals dealer houses. Brokers from commission houses, who trade for investors, are in front of me. The news headlines we watch keep us up on interest rate changes, stock news, oil prices, the state of grains, currencies and other commodities, and general news. All relate to the way we think gold prices will react.

Before I begin trading, I review the prices at which various metals closed the previous day—their settlement price. They are listed on a chart that my clerk furnishes me. I arithmetically adjust the differences between prices in the current most active trading month and the next most active trading month, arriving at a value I compare to current annualized interest rates. (Many delivery months, extending beyond one year, are listed for most futures contracts.) I also note the day-to-day differences in prices and compare them with interest-rate fluctuations the same way. For example, gold for August delivery may be priced exactly like gold for December delivery and be similarly storable, liquid and well supplied in warehouse receipts. The only difference in price, then, would usually be dominated by the costs of interest rates, storage and insurance.

I specialize in what is known as spread trading. That means I try to profit from variations in prices in two different months. It is a more sophisticated strategy than simply attempting to profit from the rise or fall of a single contract. I will trade later delivery months against the first most active futures month at any time. The month closest to delivery sees the most volume, as a rule. In fact, I specialize in executing orders between different storable metals, even if they are traded in different rings or on different exchanges of the CEC floor. I often rapidly walk about 150 feet from the platinum pit on the NYMEX to the gold on Comex to execute trades. And I walk about 16 feet between two Comex pits to execute silver/gold ratio trades.

To get these spread trades done, I have at least four risk elements to keep in mind, but the challenge is intriguing and enticing. The risks depend upon each situation. For example, in a platinum/gold spread, I must buy two 50-ounce contracts of platinum for each 100-ounce gold contract that I sell. This is risky because there are times when I am not able to acquire the two platinum contracts at the same time at the same price. A constant risk lies in liquidity. I spread between the most active month and a later month which often has very little trading. If there are not enough traders, I may find it impossible to execute the second let (part) of my order.

Another liquidity problem is the difficulty of executing certain trades in certain trading months. An August–December gold spread is easier than an August–October gold spread due to the fact that Comex is the only exchange that trades October gold.

Three Chicago exchanges and several others around the world trade gold (with less volume than Comex), and some professional traders try to profit by buying and selling between exchanges. This arbitrage helps liquidity and, conversely, when there is no possibility for arbitrage there is very little liquidity.

My spread trades are risky because of the necessity of having to move from one trading pit to the next. In the course of this physical dash, I may simply miss the opportunity to cash in on the ratio I want between two contracts.

A classic example of my problem in completing a spread occurred one day when I was in the middle of the gold ring attempting to purchase one trading month while selling a different trading month to complete a spread. As always, I hoped to profit from the differences in the sell and buy prices.

On this particular day the market started out uneventfully, but suddenly a roar went up from the crowd. Traders began to scream louder than usual trying to buy contracts in the nearest expiration month. Some of us immediately looked at the news machine in the middle of the ring, but found nothing significant on the screen. Something was happening, but we didn't know what. The commission house brokers, who normally had international telephone connections, were too busy watching their orders to talk.

Finally the denial of a rumor of Egyptian president Anwar

Sadat's assassination appeared on the video screen. We realized that something serious was going on in the Middle East. When political turmoil bubbles, we knew, people would want to fall back on gold.

The market began to rise. Local traders discounted the rumor and sold short.* Their views were supported by money markets and the stock market, which showed little change. Gold continued dropping back following denials of Sadat's assassination, but then heavy European buying reared its head, sending gold up almost $10.00. Once again, the floor traders sold contracts as the stock market and money markets remained quietly lower.

But the news services then carried stories that there had been an assassination attempt. Near the end of the trading session (Comex gold closes at 2:30 P.M., New York time), waves of buying developed. Floor traders who had short contracts were forced to buy to cover their positions. The net result, in a very stormy trading session, was a run-up in gold prices. Gold moved as much as $16 an ounce in six minutes. For the day, gold moved a total of $25.00, closing at $451.70.

That night news stories confirmed that Sadat had been murdered. Gold opened even higher the next day.

How Not to Read the Precious Metals News

Gold and the media have a unique relationship because the price of gold for many people is a signficant reflection of the economic well-being of the world. Thus, anyone who wants to pursue trading on a steady basis must understand how news that affects gold prices is reported.

All types of research publications are available to keep traders current. But it is important to understand how to interpret news through a trader's eyes. One must remember that in precious metals trading, bad news is good news and good

*For a definition of this and other terms, see Glossary.

news is bad news. That is because most short-term traders in the precious metal markets are so sophisticated—or cynical—they will interpret news very differently from the way the media presents it.

Good news, such as peace, can be deemed bearish for precious metals. High interest rates mean bad news for precious metals because the cost of carrying gold increases. Since holding gold, silver or platinum in itself means committing capital, it connotes a loss of opportunity in the interest-rate market. In other words, while you have your precious metals sitting in a vault, they produce no income, whereas their cash value could be invested in high-yielding securities.

Let's examine the whys and wherefores, the do's and don'ts of the relationship between the news and gold prices. The first rule of thumb, learned over many years on the exchange floor, is to try to interpret those who interpret and to outguess those who outguess. In other words, I try to assess the impact of certain stories carried by the media. If I find that an impact doesn't exist in the market, I will take the opposite side.

For instance, if a certain news story alleges that the mining or shipment of certain precious metals is interrupted, or that a currency will be devalued, or, for that matter, that an assassination has been attempted, I expect to see a rally. I quickly consider alternatives if the rally fails to occur. I may see that there had been an upsurge, for other reasons before the news came out, leaving few who have not already bought. I will then look for some selling opportunities in the market. Certainly, if everyone finds they have bought for the same reason and no one is left to buy, prices will tend to tumble quickly. Thus, don't be afraid to change positions often and rapidly.

Therefore, you have to watch the reaction to news. If gold has been running up, but bullish news no longer advances the price, that means a weak technical condition exists in the market. On the other hand, if gold refuses to fall despite bearish news, such as a major broker's sell recommendation or good news about inflation, I look for a buying opportunity.

I think that the media today are more influential than they have ever been. The consumer has become much more sophisticated, and demands media coverage. The futures industry also has attracted more media attention because of the great advance in the number of contracts traded. More stories than ever before are carried that have an impact on the financial community that trades precious metals.

Let's look at some fundamental ideas of news analysis. First of all, the news generally is interested only in sensationalism. One has to discount the sensational aspect of the story and relate it to the reality of what is actually being traded. You will discover, for instance, that mining companies issue statements generally bullish about their own stock if they have public shares outstanding. They also try to issue good news on consumption of the metal. Generally, news from mining companies or from journals that parrot press releases will be optimistic.

News today flows more abundantly than ever before because competing quotation services and other vendors of computer and news services are inundating the market. Remember that those who have a TELERATE, a REUTERS, a GTE or other news machines have equal access to the market, no matter where in the world they are located. They all can assess the news on the same basis. Thus I would not recommend that anyone trades precious metals unless he or she has access to some kind of news machine and quote machine to keep them up to date.

I personally feel that those who do not have physical access to these devices should not trade by themselves. Either they should always have stop orders in the pit (telling a broker to buy or sell when a price level is hit), or authorize a broker to make transactions for them on a prearranged program.

Here are some of the media tools I use daily. I read *The Wall Street Journal* every morning. There are some things I skip and some that I think are important. I read the commodities futures page in the second section, and I read the front page, which gives a marvellous synopsis of all major news covered in the paper. I read the financial section of *The New York Times* every day. I find that the *Journal* and the *Times* overlap, but don't always cover the same stories in the same way.

I buy *Barron's* every Saturday, and I read its commodity column by Richard Donnelly. I think that anyone trading commodities has to keep abreast of the market, and *Barron's* is a very good source of information for me. I read *American Metal Market*, a daily trade paper, and *Futures* magazine, an industry monthly. I read *Business Week* occasionally—if there is something about precious metals, government contracts or auto sales in it. I also occasionally read *Fortune*, but only when it covers mining companies or precious metals companies.

I also receive all the material handed out by the exchanges. I am continually checking the statistics they release—volume, prices and open interest—and I think all traders should keep abreast of these figures. They tell you what markets may be worth trading. I also read certain political and foreign affairs magazines because it is important to be aware of the economic and political climate of those countries involved in the production of precious metals.

In order to start my days, I have before me every morning a chart furnished by the exchange, which is available to anyone on a daily pickup basis. It shows all the closing prices of the precious metals from the night before, as well as changes in open interest. The opening call in London, which sets the tone for the Comex opening, and a summary of all the previous day's and early morning's major economic and political news is important to know. I also take a look at the open interest changes, the total number of unfulfilled contracts waiting for delivery, and volume from the day before so that I can relate it to what the price advance or decline has been. While trading precious metals during the day, I carefully watch the behavior of the following: the dollar against foreign currencies, the major interest rates, the grain markets, the oil and energy markets and the stock market.

Interpreting what you read is a more difficult task. Here are a few hints. Any time there is a consensus toward a particular side of the market, you can be sure that the precious metals traders will see profits to be made by taking the other side of those forecasts. For example, a major mining discovery would translate into short-term bad news because increased supply would lower prices. Traders will then analyze the new supply, figuring out if there is a cost of production involved and how important a discovery has been made. If the figure is large, they sell. If they think the news is inflated, they buy.

Major government contracts awarded in certain industries are very important to evaluating prices of industrial metals such as silver and platinum. Silver traded on Comex, is used, for example, in oxide batteries. Platinum, traded on NYMEX, is used in the catalytic converters of automobiles. A large government contract to manufacture, say, military vehicles, naturally translates into higher prices for these precious metals.

Consider the balance-of-power figures. Anytime there is a surplus in trade figures for the U.S., that means bad news for the precious metals market. Also, strikes, labor problems and government dislocations are vital events for the traders to follow. If miners go on strike, for example, there could be an interruption in the delivery process leading to an increase precious metals prices.

Good news to most people consists of an up-move in the gross national product. This often means an increase in personal disposable income, a rise in capital spending, lower unemployment figures, and increases in sales of automobiles, machine tools, housing and construction. But gold traders know that precious metals drop on such news, and act accordingly. When the dollar is booming, people don't want gold. Other news that the public cheers, but which causes traders to sell, is: downward movement of the cost of money (lower interest rates), a lower consumer price index, less inflation, lower wholesale prices, and a slowdown in the growth rate of the U. S. money supply. What should be bullish for most is bearish for us.(For many years the popular misconception of lower interest rates being bullish for metals hovered over traders' heads. The $800 gold price was broken at the 21% interest rate, and the subsequent 11% interest rate has helped gold decline to below $350. It is only the inflationary *expectation* that is bullish.)

Once a week the Federal Reserve releases its money supply figures. When there is an increase in the money supply, floor traders believe the Federal Reserve will have to tighten interest rates. This translates into a downward price movement for precious metals. Traders who have positions in anticipation of news and who try to offset them before the news is released send the prices into the reverse.

Traders are less unanimous about interpreting the stock market. Upturns in the market will be seen as bullish by some and bearish by others. Those who say it is bullish do so because they think a rising stock market means more disposable income for investors who are willing to spend more on precious metals. Those who take the other side believe that if the stock market is going up, who needs precious metals?

When the CPI (Consumer Price Index) rises, it is an inflationary signal. Those who like precious metals as a hedge against inflation say a CPI jump is a time to buy precious metals. I agree. If there is going to be inflation, I'd rather own precious metals than any other form of investment.

Let's take a closer look at another vital source of information: the daily reports issued by exchanges. These give open interest, contract price and volume, or number of contracts traded. Those who regularly trade on the floor of an exchange know, for example, that an increase in price accompanied by an increase in open interest and volume is bullish.

On the other hand, a decline in price, open interest and volume is normally an indication of people liquidating their positions. If prices decline sharply while volume and open interest increases, this means that there is very heavy short selling, which later could set a bullish stage for a big move up. Short selling is the sale for future delivery of contracts not now owned. The seller hopes to repurchase the contracts for delivery at a future date at a cheaper price.

One must also remember a very crucial point in interpreting daily news—it gets old very quickly. Once the markets have adjusted to a particular piece of news, they will no longer react. I call this the "diminishing value" syndrome. For example, reports of a war frequently lead to an initial burst in precious metal prices. But, as time goes by, the rallies get smaller and smaller. For a short while everyone has bought up to their ears and is looking for an excuse to sell. The stage is set for a sharp downturn.

Let me now give you a news alert. Avoid getting financial news only from general radio or television stations. They carry four times the amount of sports news as financial news. They regularly make mistakes in reporting precious metals prices—for example, the London price in the morning. Also avoid reading reports that seem based on sensationalism. Instead, concentrate on a news page that seems to have some regular credence and authority.

One thing I have always found to be a good idea in trading is to be a "contrarian." Therefore, I try to know what the major brokerage firms are recommending and if they are all on the same side of the market.

If I discover a trend, I will look for evidence that the brokerage firms have built up their positions. Then I will look for the beginnings of the large contract positions to crumble, as clients try to capture profits. The fact that a brokerage firm has already finished doing a lot of buying, certainly means that other sellers cannot be far behind. Once they have all finished buying, and there is nothing else left to buy, the rush to get out begins slowly and builds up.

The computer services that offer technical trading programs usually have computer-generated "stop orders" in the market, meaning that at certain levels they want to buy and at certain levels they want to sell. It has been my experience that computer services like to sell at a new high in the market and buy at a new low. Nimble traders anticipate this.

Some Regularly Reported Indicators and What They Usually Signify to a Commodities Trader*

Indicator		Signification
Agricultural prices—higher.	⬇	Inflationary. Purchasing power of the dollar decreases.
Consumer Price Index—higher.	⬇	Inflationary.
Gross National Product (the total amount of goods and services produced)—lower.	⬆	Shows a slowing down in the economy, signalling that the Federal Reserve Board will loosen money by allowing rates to come down.
Housing Starts (new housing construction)—higher.	⬇	Shows a growth in the economy, and, with new housing demand, prices rise and mortgage rates rise. The Fed is less accommodating and attempts to tighten by allowing rates to rise.
Industrial production—lower.	⬆	Indicates a slowing in economic growth, signalling that the Fed will be more accommodating and allow interest rates to drop.
Money supply figures—higher. (M1-A equals cash plus regular demand deposits; M1-B equals M1-A plus checking-type deposits.)	⬇	Excess growth in money causes inflation, generating fears that the Fed will tighten money growth and allow short-term interest rates to rise. This will cause lower prices in the bond market.
Personal income—higher.	⬇	Inflationary. An indicator that consumers will buy more, leading to higher prices to meet demands.

Retail sales—higher.	⬇	High retail sales are an indication of economic growth, meaning the Fed will be less accommodating.
Unemployment figures—higher.	⬆	High unemployment indicates a lack of expansion within the economy, a positive sign for the bond market.
Wholesale Price Index—up.	⬇	An inflationary indicator, because as demand for goods rises, prices also rise. This means the Fed will be less accommodating in allowing rates to go lower.
Fed buying bills.	⬆	When the Fed adds money to the system, interest rates lower, due to the feeling that the growth of money is not expanding at an accelerated rate.
Fed is tightening.	⬇	The Fed, fearing excess growth in the money supply, decides to allow Fed funds to rise, causing other short-term interest rates to follow upwards.
Fed raises the discount rate (the rate at which member banks of the Federal Reserve can borrow from the discount window at the Fed).	⬇	An increase in interest rates between banks and the Fed means an increase in rates to customers will follow. The Fed expects the measure to slow down inflation.

Fed does Repos (reposit).	⬆	Fed puts maey in the system by purchasing collateral and agreeing to resell it at a specified time. Increase in the money supply brings rates down.
Fed does either reverse or matched sales.	⬇	Fed takes money from the system by selling collateral and agreeing to repurchase it at a specified time. Decrease in money supply pushes rates up.

* ⬆ = Commodity prices should rise.
⬇ = Commodity prices should fall.

TWO

What Happens to an Order

Brokerage firms and banks are the major initiators of transactions on commodity exchanges. Some brokerage firms are active in both securities and commodities. Some are specialists in commodities only.

Floor brokers are traders who execute the brokerage firms' orders on the floor of an exchange. Some floor brokers also trade for their own accounts, but rules require them to fulfill customers' orders before they act for themselves. Some floor brokers are independent and may act for anyone. Some are employed by brokerage firms and might be under their employer's order to trade only for the firm.

The investors and speculators are a major part of any trading market. They make markets so that users of a commodity can transact their purposes—to buy a commodity or to hedge their inventory market. A private investor may act for himself or herself or may be advised by an account executive, investment advisor, or commodity pool operator. Though most private investors use brokerage houses to carry out their trades, some select their own floor brokers on exchange floors. The floor broker will then execute orders, and clear them through the investor's designated brokerage firm.

On most exchanges, commodity-only brokerage house employees make up about 25 percent of the membership, trading house bullion dealers and physical traders make up another 25 percent and independent traders (called locals) and brokerage firms dealing in commodities, the balance. Commercial firms use the futures markets to hedge business they do in the physicals and other markets. Most brokerage firms do a large amount of retail and speculative business.

When a trade house or professional trader makes transactions on two different exchanges at the same time, it is called arbitrage. Arbitrage is the near simultaneous buying in one market and selling the same commodity (and same delivery month) in another market, at a differential that would insure a profit. Arbitrageurs zero in on slight price variations that quickly surface on different exchanges.

In addition to understanding the market participants, it is necessary to understand trading practices.

The members of Comex wear badges with letters and numbers on them because there are so many members that buyers and sellers might not recognize each other, or take the time to look, in the heat of trading. (The letters frequently are a trader's intials). Traders cannot depend on spelling each others names correctly. The system of trading on the Comex is known as recognition trading.

The system of trading on NYMEX is known as pit card trading. In pit card trading, the seller fills out a card with the buyer's name or numbers. (NYMEX names also are the abbreviations or four-letter code words that identify a particular member.) The seller hands in the pit card with the seller's code name on the top of the card, the contract month and number of contracts. The seller also puts down the buyer's code name and price of the transaction. This card is handed to—actually thrown at— a "pit reporter" stationed in the center of the ring. His or her job is to receive these pit cards. The pit cards are collected, time clocked and then keypunched into a computer system. The goal is to give some semblance of order within a one-minute time frame (the time clock does not show seconds) as to who purchased how many contracts of what particular month from which seller.

Many people mistakenly believe that this system gives a very good sequence of the times in which sales take place. Actually, in busy markets nothing could be further from the truth. If a broker on the left side of the reporter throws in the pit card before a broker on the right throws, the left sale may be recorded first, even though the broker on the right may have executed a trade just as fast. Also, he or she might have had to put in three transactions on the same card while the first broker had only one trade.

All trading on all futures exchanges in the U.S. happens by open outcry. That means a trader cannot quietly sell lots to the person standing to his or her left or right at a price known only to the two of them. The trader must offer his or her bid or ask in a loud tone of voice so that any broker in the ring may participate.

In the heat of trading it is very common for brokers to write the wrong number of contracts, the wrong prices and/or the wrong names on their trading cards. This results in what we call "out" trades, or errors. While most brokerage firms and texts tell you that out trades and errors are a minuscule part of the business, in my ten years of experience I have found that in very busy markets out trades can be as much as 10–20 percent of the gross value of the brokerage business transacted by a busy broker in any one year. (Brokers on the floor are usually responsible for settling errors among themselves. Customers are rarely affected.)

Futures contracts are standardized among most exchanges for liquidity and for purposes of arbitrage.

Lawrence J. Bilello is a member of, and floor broker on, the New York Mercantile Exchange and the Commodity Exchange. He joined the floor in 1975.

How does somebody become a member of an exchange?

LB: You must buy a membership and you must buy it from one of the current members. You don't buy a membership from the exchange, but from some other member who is selling or leasing a seat. (The only time you can buy directly from an exchange is when a new category of limited membership is created.)

Let's take a look at the floor population. Who are the members of the exchange who trade in the ring?

LB: There are some who are called floor brokers and there are some who are called floor traders. Floor brokers are those who execute orders for commission houses. Floor traders, sometimes known as locals, execute trades for themselves.

All members have a badge with numbers and letters on it. Does that mean that you write down the numbers and letters of everyone you trade with?

LB: As a matter of expedience, because the clearing system that reconciles the trades at the end of every trading day goes strictly by a person's number on the Commodity Exchange, most members are identified to me by number. Therefore, on my trading cards I identify them by number only, their trading badge number.

If you were opening an account tomorrow and wanted to enter your own trading orders, how would you pick a floor broker to handle your orders?

LB: I would do certain investigations myself, just as when I pick a brokerage firm. I would look at the reputation of the broker. I would be concerned about the level and quality of exeuctions—the performance level, how fast my order was filled and at what price. I'd probably want to interview my own floor broker. I'd ask certain questions about how he or she would handle particular orders.

Some of the questions I would ask are whether or not the broker trades for himself or herself. I believe that it is very important to trade for oneself, as this makes one more cognizant of the market. Since this is a futures market, it is necessary not only to be aware of where the liquid month is trading (with the greatest amount of contracts being sold in this liquid month), but also all of the following months and their relative values.

The next question I would ask would be the kind of floor operation the broker has. Is it one ring [pit] such as gold, or multi-rings such as silver or platinum? The latter two have an inherent relationship to gold, while heating oil, which is an inflationary market, also has an underlying relationship to gold. I prefer someone who trades in multi-rings, as this is important for overall experience. For instance, if I put in an order to sell some gold contracts and the broker sees that silver is rallying, the chances are very good that the gold will strengthen and the broker will be able to scale up my order and get a better price for me. The broker should never be the only seller in a market which is going higher.

All of the above does not guarantee a riskless situation, but it does show a greater depth of floor and trading experience. The best way to get that experience is by trading your own account. It is very easy to tell someone to do something. It is a lot harder to take that risk yourself.

What sort of customer orders do you execute and who are your customers?

LB: Quite often I don't know who the customers are. Orders are transmitted to me from a customer's registered representative at a brokerage house. The registered representative handling the customer's order gives it to a phone clerk in the brokerage house office who transmits the order to the floor of the exchange. A telephone clerk at the exchange receives the telephone order and brings it to me in the pit. All I see is the order with the brokerage firm code.

How does the order get to the ring? Does the runner hand you the order?

LB: Quite often the runner doesn't. If it is a market order, which means it must be done immediately at the best price available, it *is* handed to me immediately by the runner. But if it is a limit order, asking to buy or sell at a certain price, then it is grouped by the clerk with a stack of other orders. This stack represents our "book," similar to a specialist book on a stock exchange. Since the order is not, in fact, a valid order because it's not at the prevailing market price, it's held with the other bids until such time as the market settles down to that particular area.

You mean, for example, if a customer puts in an order to buy a futures contract of gold at $398.00 per ounce, and the market is currently trading at $400.00, that order is not handed to you. Instead, it is grouped with orders to buy at $399.50, $399, $398.50 and $398.00, all of which may be in your hand?

LB: Yes, at all times. But often I will have an outside clerk who would maintain what we call "the book" with the best bid and offer to buy or sell under or above the market, filed according to time precedence. It's similar to a specialist book on the American or New York Stock Exchange.

Suppose two customers from two different commission houses wish to buy gold at $399.00, and suddenly the price hits the mark. How do you determine which customer makes the $399.00 buy?

LB: By the purest system available. The first order in by time stamp gets priority as far as similar bids or offers.

During the course of a day, you often handle thousands of orders. Let us say that you have a lot of buy orders and a lot of sell orders all at once. How do you know which orders to do first?

LB: If there is an inordinate number of buy orders paying higher prices, those customers must take precedence over any other consideration. I refuse to be the only floor broker who's a seller, so what I will do for my customers is buy contracts until I can match off prevailing sell orders at a fair price.

Larry, this is complicated, so let's try it from another angle. Let us say that you have ten different orders to buy and ten different orders to sell, all to be done at the market. How do you know which order to execute first?

LB: I have no opinion or judgment of the market, so I will match the orders off at a prevailing price in the market that I feel is equitable. At one particular time, I search for the best bid or offer. I will try to meet the middle of the bid or offer. Quite often in a market that is moving very quickly there are certain gaps. By taking advantage of this, I can do justice to my sell orders as well as to my buy orders.

Very often a customer doesn't understand what has happened to a particular order. He or she may be sitting in a branch office of a member firm, or watching the prices through a computerized service, and see that though gold is trading at $399, his or her buy order at $399 has not been filled. How would you explain to them your inability to buy, even though prices on the tape said $399?

LB: All my orders are on a time preference basis, so the first orders in at that particular bid have preference. The second thing the customer has to realize is that the commodity ring is a true auction market. His or her bid or offer is not necessarily the only bid or offer at a particular time and at a particular price.

So now what you're telling me is there may have been a lot of brokers bidding at $399 for their customers. Since very few contracts may have traded at that price, your customer did not get one of those contracts because you were unable to participate in a transaction?

LB: Yes, or I may have had a time-stamped order two minutes before a particular customer's order. Also, I may have had twenty orders left to buy ahead of this particular customer's one-lot limit bid order. The rules of the exchange say I must buy all twenty contracts by time preference before I can buy one contract for my customer.

Tell me a little bit about how crowded, or confusing, it gets in a trading pit; on Comex, for example.

LB: Comex has over 770 members, but out of that number I venture to say only about 160 to 180 congregate in any one particular ring. During the trading day we have an unwritten rule that says that no one goes to lunch, no one leaves the ring except for calls of nature.

Does it matter where a broker stands in the ring? Are there better or worse spots?

LB: We try to stand as close to the center of action as possible, and we also stand as close to our telephone stations or booths as possible so that our clerks can rapidly reach us through a thick crowd. The idea is to hear *all* the bids and offers. A broker also has to be as loud as he or she possibly can.

So, voice level is also very, very important.

LB: Yes, as I said before, a commodity exchange is a true auction market, and all bids or offers come via open outcry. I have taken voice training courses. Most traders who project well have studied either the technique of opera or some other type of voice projection.

With so many brokers all yelling and screaming at the same time, all trying to do orders for a lot of customers and commission houses, do you have errors, what we call "out" trades, when you buy or sell contracts? And how do they affect a customer?

LB: We have out trades, especially in volatile markets. But the customer is never affected by an out trade.

You mean that if you tell your clerk that your customer has purchased something at a particular price, and later you find that through some error made by you or the party you bought from you were unable to purchase that contract at that price, the customer still receives the execution?

LB: That's correct. There's a code of ethics on commodity exchanges that obliges me to fulfill an obligation to the customer. I personally might take half or more of the loss.

A typical misunderstanding could be that you may have an order to buy nineteen contracts and you shout across the ring. From across the ring the other broker sells you what you think are nineteen contracts, but you find out later that he sold you only nine contracts. That means that you are unfilled on ten contracts. If the market is now higher than the price that your contract called for, do you tell your clerk to report to the customer that you've purchased the nineteen contracts?

LB: Yes, my understanding was that nineteen contracts were purchased for my customer, so any losses or any misunderstandings are settled between the two brokers.

Do out trades happen very often?

LB: The out trade rate in normal markets is roughly two percent, and in a market which has the volatility potential that gold or silver or any of the commodities have, that's a commendable record.

*What are the most popular types of orders you see?**

LB: The "limit order" is very important. It is a buy or sell order at a certain price. It cannot be technically executed at a price above or below the limit that the customer sets. However, there are certain circumstances where, in the heat of a very volatile market, bids and offers may be violated. But a basic limit order definition is an order to buy or sell a contract or an option at a specified price or better.

*For a summary of orders a customer can make, see pages 43-45.

The orders are for the day, unless they are specified for the week, or as GTC (good till cancelled), which is an open order.

What is a market order?

LB: An offer or bid to take the prevailing offer or bid in the market. A market order is executed within seconds of the time it reaches the trading ring.

What is a stop order?

LB: A stop order is a very important tool. It is an order that becomes a market order when the market reaches a certain level. Sell stops are entered below the market; buy stops are entered above the market.

Why would somebody use a stop order?

LB: For protection. It's used at critical chart points, or technical points, in a market to "stop into" a position.

So if somebody owned gold at $395 and the market was trading at $400, to protect against price deterioration, one would enter a stop order to sell at $395? That means the order would not become elected until the market trades back down to $395. Is that right?

LB: That's correct. That order, when elected becomes a market order and must compete with available market orders in the floor broker's hand as well as with other stops that are elected and become market orders. There's no guarantee that the prevailing market will allow a floor broker to sell at $395. There may be no offers to buy at that price and the order will be sold at the best price—which could be lower.

Let's look at stop–close orders.

LB: "Stop–close only" is a technical order. It initiates or liquidates a position during the closing period of certain markets. It's only executable during the closing period of a futures market, and it can be elected from either side, buy or sell. In other words, a "stop–close only" order becomes a stop order if that price is reached at the close of the day at the point of trading. There is an important point to be made here. If a market is declining rapidly on the close, there may be certain "stop–close only" orders that are elected at the last tick of the day. (A tick is a price move up or down.) It is physically (and legally) impossible to liquidate a position *after* the market has closed. So, though a last tick may elect the "stop–close only" and become a market order, it cannot be executed on the last tick.

What is the difference between an "or better" and a "not held" order?

LB: "Or better" is a limit order with a little discretion; however, the limit may not be breached. "Not held" is an order with more discretion, but the broker is not held liable for not participating at different price levels. As far as floor brokers are concerned, the difference is that the broker has market judgment discretion. The broker is entitled to pull back on particular bids or pull back on particular selling to try and do better for the customer.

What is a "trailing" order?

LB: A trailing order is a market order that's not executable unless the market has risen to a specified amount in case of a sell order, or fallen to a specified amount in case of a buy order. But it must be executed at whatever price is achieved. Trailing stops are automatically adjusted; limits are rigid.

What is a "stop limit" order?

LB: A stop limit order is probably the most misunderstood order of all. It is an order to buy or sell, with buy orders entered above the market and sell orders entered below the market. It is similar to a regular stop, which we previously discussed, except that there is a limit placed by the customer.

Are there large pools of commodity orders that are computer-directed, that have stop orders which sometimes cause a rapid movement in price?

LB: Yes, quite often what will happen is that markets will reach a certain sensitive area where stops are elected and become market orders. We call those periods "breakouts" and "breakdowns" in the regular market.

Let's get down to some floor mechanics. Can anyone execute an order on the floor of a commodity exchange?

LB: Only a bona fide floor member of an exchange can execute an order.

Does the broker have to be registered?

LB: Yes, anyone doing any type of customer business must register with the Commodity Futures Trading Commission (the regulatory agency overseeing the futures industry).

What do you do to prepare for your trading day?

LB: I am in constant contact with all the news services at night and also in the morning before I come into New York. (I have a two-hour daily commute.) As far as preparation, it is mainly mental. I call my contacts worldwide prior to the opening to ascertain exactly what has occurred on the other markets in different time zones, who were the large players and what were their reasons for their trades.

Types of Orders

1. *MARKET ORDER*—A market order is an order to buy or sell a stated number of futures contracts or futures options at the best price obtainable immediately after the order is transmitted to the ring.

2. *LIMIT ORDER*—A limit order is an order to buy or sell a stated number of futures contracts or futures options at a price not less favorable than that specified in the order.

3. *SPECIFIED TIME ORDER*—A specified time order is an order to buy or sell a stated number of futures contracts or futures options which, if not executed within the time specified in the order, automatically expires. A specified time order includes, without limitation, (a) a fill-or-kill order, (b) an order executable only on the opening or during the opening call, or (c) an order executable only during the close.

   EXAMPLE:

   A. *Fill or Kill Order (FOK)*—This order is to be executed immediately at the time it is entered. The broker must attmept three times by open outcry to execute it. If it cannot be executed, the order is automatically cancelled.
   B. *Opening Only*—this is an order that is executable only on the opening.
   C. *Close Only*—This is an order that is executable only on the close. If a market is declining rapidly on the close, there may be certain "stop–close only" orders that are elected at the last tick of the day. (A tick is a price move up or down.) It is physically and legally impossible to liquidate a position *after* the market has closed. So, though a last tick may elect the "stop–close only" and become a market order, it cannot be executed on the last tick.

4. *STOP ORDER*—A stop order to buy becomes a market order when a transaction in the futures contract or futures

option occurs at or above the specified price after the order is received in the ring. Also, if specified by the customer, a stop order becomes a market order when a bid for the futures contract or futures option is made at, or above, the specified price. A stop order to sell becomes a market order when a transaction in the futures contract or futures option occurs at or below the specified price after the order is received in the ring or, if so specified by the customer, when an offer for the futures contract or futures option is made at or below the specified price.

EXAMPLE:

Customer Long 1 Dec Gold 64800 (a gold contract bought for December delivery at $648).

Market Trading 65000 (the contract is currently trading at $650).

The customer feels the market is going higher but, in case he or she is wrong, wants to protect part of the profits. The customer would put in a sell stop order at 64900. If the market should start going lower, once it touched 64900 the order would automatically become a market order and should be executed somewhere around 64900.

Exception—Erratic Markets

Sometimes markets have large swings. In a case like this, a customer who uses stop orders could get hurt. The market could trade at 64900 with the next trade at 647.50, and the lower figure could wind up as the execution.

5. *A STOP LIMIT ORDER*—A stop limit order to buy becomes a limit order when, after the order is received in the ring, a transation in the futures contract or futures option occurs at or above the specified price or, if so specified by the customer, when a bid for the futures contract or futures option is made at or above the specified price. A stop limit order to sell becomes a limit order executable at the limit price when, after the order is received in the ring, a

transaction in the futures contract or futures option occurs at or below the stop price. Also, if so specified by the customer, it became executable as a limit when an offer for the futures contract or futures option is made at or below the specified price.

6. *MARKET-IF-TOUCHED ORDER*—A market-if-touched order is the same as a limit order except: (1) A market-if-touched order to buy becomes a market order when, after the order is received in the ring, a transaction in the futures contract or futures option occurs at or below the specified price; (2) A market-if-touched order to sell becomes a market order when, after the order is received in the ring, a transaction in the futures contract or the futures option occurs at or above the specified price.
7. *TIME-AND-PRICE DISCRETION ORDER*—A time-and-price discretion order is a market order or a limit order to buy or sell a stated number of futures contracts or futures options in which a floor trader is to exercise his or her own judgment and discretion in determining the price and the time of execution of the order.
8. *NOT HELD ORDERS*—A member may attempt to execute an order not held on the exchange, which means the broker may exercise discretion to some extent. He or she will not be held liable by the customer for not participating at different price levels.
9. *CANCEL FORMER ORDER*—This order cancels a previous order.
10. *ONE CANCELS OTHER (OCO)*—This order cancels a previous order and replaces it with a new order. Can be used with Day or Good Till Cancelled (GTC) orders.
11. *GOOD TILL CANCELLED (GTC)*—This is an order that will remain open beyond the usual day limit until it is specifically cancelled.

THREE

A Brief History of Gold Futures

A series of dramatic events set the stage for the sharp fluctuations in the price of gold— "gold fever" —that began in 1972.

That year's devaluation of the dollar, the leading currency in the world, prompted international monetary sophisticates to ponder the possibilities of gold and gold-related investments. Soon after, in 1973, the first Arab oil embargo shocked the west and stocks plummeted. People began looking to precious metals as an investment for an uncertain time.

As of January 1, 1975, U. S. citizens have been permitted by Congress to own gold. This changed the entire complexion of the world gold market. International traders began increasing their inventories in anticipation of a "new American buying wave." However, Americans had been unaccustomed for the previous forty years to direct investment in gold, gold futures or coins, so the buying wave never materialized. Gold started trading in the U.S. at about $200 an ounce. It had been fixed at $35 an ounce since 1934. The price dropped quickly from the $200.

In their rush to reduce inventories of unwanted bullion, dealers reduced the prices down to a little over the $100 level through 1975. (Dealers do not like to hold excess merchandise at any time because the capital tied up in gold and storage could be earning interest income.)

There was a general fluctuation in the price of gold from early 1975 to 1976, influenced by the first U.S. gold auction. Between 1975 and the end of 1976, when the International Monetary Fund decided to de-monetize the precious metal, gold dropped sharply to below $120 an ounce from a little over $200 an ounce.

The market had been extraordinarily quiet when gold futures first started trading, with prices generally lower until September, 1976. It was so slow that some traders played chess in the gold ring at the Commodity Exchange in New York, the leading gold futures exchange in the U.S. There was little else to do do. However, prices finally started to rise to over $165 in late 1976, after a decline to about $100, and trading volume of gold futures contracts moved up to about 100,000 contracts per month in 1976–1977.

Since then, dramatic increases in both price and volume have occurred. For example, in December, 1980, more than one million gold futures contracts (at 100 troy ounces per contract) were traded in a single month. That's 100 million ounces of gold!

This was not just what professionals call "in-and-out trading" —buying and selling for quick profit. Rather, open interest, which is the total number of unsatisfied futures contracts that are held, was rising—a sign that people were trading with delivery in mind. The total open interest climbed from insignificant levels at the end of January, 1980, to 250,000 contracts by the end of December.

By fall of 1977, 100-ounce gold bars, the quantity traded on the Comex, started to flow into exchange-approved warehouses in the United States. This is, of course, not surprising, as physical stocks are always needed to support trading. Obviously, when so large an open interest exists, it would be possible without adequate quantity of deliverable stocks to create a "squeeze" situation. This could happen if people who have purchased the gold for future delivery at a certain price demand delivery on one of the contract maturity dates from "goldless" shorts (those who have sold at a given price for future delivery, and who must repurchase their contracts or deliver by a certain time). There must be enough deliverable gold in the exchange-approved warehouses to prevent such a squeeze.

A squeeze can be perpetrated by longs, people who buy futures contracts and/or physicals (actual bullion) in such great quantities as to cause a shortage. This will force up the price, so that the shorts (the people who sold in anticipation of a lower price) will have to increase their amount of committed capital (called margin, which is a good-faith deposit) or repurchase previously sold contracts at higher prices.

If there is a political or economic event to support the buying, then the shorts will encounter great difficulty in obtaining the required bullion.

Gold and the Related Metals Markets

Since 1666, when King Charles II gave London bullion merchants control over gold and silver dealings, British trading activity has been centered in London. The Empire's large gold discoveries in South Africa and Australia made London the center of world gold trading. It remains so today.

There are five major firms in the physical (actual bullion) gold market, all of which have been dealing for 100 years or more. One of them, Mocatta and Goldsmid, the oldest British house, was founded in 1684, ten years before the Bank of England was established. The four other major London market participants who set the London morning and London afternoon "fix" every day are Samuel Montagu & Cie, Sharps, Pixley & Cie, N.M Rothschild & Sons, and Johnson Matthey Bankers.

Representatives of each firm meet twice a day for the so-called fixings. They sit in a room at N. M. Rothschild, where each has a table with a telephone and a small Union Jack. The flag is used for signalling price decisions. Each representative determines the amount of gold his institution would like to buy or sell, and at what price. When the fixing, or the price agreement session, begins, the chairperson calls out a figure and the representatives respond by pointing their flags either up or down, indicating whether they would buy or sell at the suggested price. If everybody wants to sell, the chairperson lowers the figure until some begin to show a buy interest at a particular level. And if they all want to be buyers, each reveals how much he would like to purchase or sell.

If there is no agreement, the chairperson, traditionally the Rothschild representative, makes a proposal to bring the figures into line. Throughout the session, all five people are in telephone contact with their trading rooms, and of course these trading rooms are in contact with other client banks and with exchanges that are open and trading.

When the fixing figure is finalized, it is transmitted throughout the world by electronic news services, telephone, telex and the next edition of many newspapers. But for a trader, next edition's prices are like yesterday's newspapers: little use to anyone but historians. Although no dealer is bound by the London gold fixing, it is a very good reflection of supply and demand in the world market and an important indicator to all market participants all over the world.

The London fix in the morning, coupled with the Comex closing price the previous night, give a very good reading on the level of interest and activity that can be expected for the coming day's trading. It does not necessarily mean that New York will open or close where London has come in. Comex gold might close at $375 an ounce on a Friday night, but prices could be affected by trading in Hong Kong late Friday night. Thus, the Monday morning gold fix from London can be higher or lower than the previous Friday's Comex price.

Despite the prominence of London as a gold center, it probably is losing some of its influence. The United States has really become the world's leader in transacting gold business.

(One sign of this is that since 1968 the London market has quoted gold in U. S. dollars per ounce.)

An unofficial major gold market is in Zurich, Switzerland. Zurich opens quickly in the morning and continues to trade through the major fixing banks in Switzerland. Because the Swiss fixings are unofficial, they are called the "kerb" (curb) market.

South Africa has been channeling the major part of its transactions through the newly established Zurich gold market. Combined with the resources of Credit Suisse, Swiss Bank Corporation and Union Bank of Switzerland, the kerb has become a major precious metal center. In 1972, when the U.S.S.R. brought sizable amounts of gold to the market, a Zurich bank handled a major portion of the sales.

Business has shifted back and forth between London and Zurich over the past few years, and both centers are highly respected for their professional expertise. In fact, most Swiss banks have a stable reputation as conservative handlers of precious metals, and they hire expert analysts in all their research operations.

New participants in the U.S. gold market are certain major American futures commission houses that do physical bullion trading. Some of these operations today allow trading on a 24-hour basis. Transactions are often offset on Comex in New York through the use of a vehicle known as "exchange for physical" (EFP). For our purposes, we will define an EFP as someone who is willing to trade a physical gold contract for a futures position.

The North American futures markets are highly liquid. Also, they are one of the few major markets where large quantities of metal can be transacted swiftly and with complete assurance to the participant that political instability will not result in a possible loss.

The Far Eastern markets are important. Hong Kong and Singapore do trade quite a bit of gold but they don't provide the same service as the American futures markets.

Recently the Chicago Mercantile Exchange began a link

with the Singapore International Monetary Exchange that allows for gold and other contracts to be bought and sold on either exchange. Such round-the-clock trading may be copied by other exchanges, including Comex, and the Sydney Futures Exchange.

The open and free flow of gold has been restrained in many nations. Certain governments have held back the free flow of gold, and a few major countries are highly restrictive. In these cases, a black market—where gold commands high prices—arises. Most of the Communist world fits this category. There are also so-called gray markets, such as some Asian, African and South American countries, where gold ownership and trading are restricted to a high degree and very heavily taxed.

In countries with a free or open market, such as ours, there are reporting requirements only. Merely a handful of nations permit the flow and ownership of gold without any inhibitions and prohibitions. It is interesting to note that anyone may purchase or sell gold or make and take delivery of a futures contract through all major brokerage firm branches in most major cities of the free world by depositing U.S. dollars or withdrawing them. To me this is a freely convertible gold standard vis-a-vis the U.S. dollar. Someday the media will discover this and stop asking about the gold standard for political purposes.

FOUR

The Major U.S. Markets

Gold

The trading hours in New York on Comex are 9 A.M. to 2:30 P.M., New York time. The trading symbol on our quote machines is GC. The trading unit and grade of all Comex gold is 100 ounces (5% more or less) of refined gold assayed at 99.5 fineness. It can be cast either in one bar or in three one-kilogram bars, and it must bear the serial number and stamp of a refiner approved by Comex. A list of approved refiners may be obtained by writing to the exchange.

Trading for Comex occurs for the next calendar month, the next two calendar months and any February, April, June, August, or October and December falling within a 23-month period. Delivery of gold bullion against futures contracts at the Comex are made at the seller's option during any business day within the month specified in the contract. Bullion must be made available to the buyer at one of the vaults licensed by Comex.

Licensed depositories are vaults judged by Comex to be safe and sufficiently well guarded to deter theft. Periodic random inspections are conducted by independent auditors at these vaults to check on the accuracy of depository receipts.

Price changes on Comex gold futures occur in multiples of 10¢ per troy ounce, which is equivalent to $10 per contract. A fluctuation of $1 is equivalent to the change of $100 per contract. During any one market day, price fluctuations of gold is limited to $25 per contract—that is, above or below the price established as the settlement price at the end of the preceding business day.

If the settlement price for any contract month hits the daily price fluctuation limit on two consecutive days, a system of expanded price limits becomes effective automatically. Price limitations do not apply to prices for the current delivery month.

Investors don't pay the full price when they buy a futures contract. They put up margin. Original margin requirements for Comex trading are designed to maintain the financial integrity of each futures contract. Margin levels are determined by an exchange's board of governors, and reflect price volatility. A trader must add to the margin on a predetermined schedule whenever the market prices go against the contract he or she holds. Conversely, a trader may withdraw excess equity as the market moves in his or her favor.

Silver

The Comex silver futures contract calls for delivery of a specific grade of refined silver in standard bars during one of the specified futures months. Trading hours are 9:05 A.M. to 2:25 P.M., New York time. The trading symbol of the quotation machines is "SI," the contract calls for delivery of 5,000 troy ounces (6% more or less) of refined silver in bars cast in weights of 1,000 to 1,100 ounces each. The silver must be assayed at a high level of fineness (999) and must be approved and listed by Comex. A list of approved assayers and refiners may be obtained on request from any silver-trading exchange.

The trading months are January, March, May, July, September and December for a twenty-three month period. Delivery of silver bullion against futures contracts traded on the Commodity Exchange are made at the seller's option during any business day within the contract month. Bullion must be made available to the buyer at one of the vaults licensed by Comex. A list of licensed depositories is available by writing to the exchange. Licensed depositories are the same as for gold.

Price multiples are registered in multiples of one-tenth of 1¢ per troy ounce, an equivalent of $5 per contract. A fluctuation of 1¢ in silver is therefore the equivalent of $50 per contract. During any one market day, prices for a delivery month can move no more than 50¢ an ounce above or below the settlement price established at the close of the preceding business day.

If the change in the price for any contract month hits the daily price limit on two consecutive days, a system of expanded price limits becomes effective automatically. Price limitations do not apply to prices for the current delivery month.

Margin requirements are on the same basis as for gold.

Copper and Aluminum

Copper futures trade on the Comex from 9:50 A.M. to 2:00 P.M., New York time. The trading symbol on the quotation machines is "CU." For every contract of copper, a seller must deliver 25,000 pounds (2% more or less) of electrolitic lake or certain types of refined copper. All deliverable copper must conform in size, shape and chemical analysis to specifications established by the American Society for Testing and Materials in particular shape and grade involved.

Copper is traded for the current calendar month, the next two calendar months and any January, March, May, July, September or December falling within the twenty-three month period beginning with the current month.

Deliveries of copper against futures contracts traded on Comex are made at the seller's option on any business day during the months specified in the contract. Copper must be made available to the buyer at one of the warehouses licensed by Comex. A list of warehouses may be obtained by writing to Comex.

Price changes in copper occur in multiples of one-twentieth of one cent (5/100), equivalent to $12.50 per contract. Therefore, a fluctuation of 1¢ is the equivalent of $250 per full 25,000-lb. contract. This is different from gold and silver.

Daily price fluctuations for each copper delivery month are limited to 5¢ per pound above or below the settlement price established at the close of the preceding day. If the change in the settlement price for any contract month equals the daily price fluctuation limit on two consecutive days, a system of expanded prices is triggered. Margin conditions are the same as for other metals previously discussed.

Comex, which calls itself "the world's most active metals market," has a new contract on aluminum. It started trading on December 8, 1983.

The Comex futures are traded in contract units of 40,000 pounds of primary aluminum (in one of two specific grades). Trading months include the current month, plus the next two months and any January, March, May, July, September and December over a twenty-three month period. Trading hours are 9:30 A.M. to 2:15 P.M., New York time. The trading unit is 40,000 pounds (2% more or less) of virgin primary aluminum. Minimum price fluctuation will be 5/100 of 1¢ per pound, or $20 per contract. Price limits are 5¢ per pound above or below the previous day's settlement price. Delivery will be at the seller's option during any business day within a contract expiration period which begins with the first notice day. Aluminum must be made available to the buyer at warehouses licensed or designated by Comex.

Precious Metals Traded on the New York Mercantile Exchange

PLATINUM

The contract for platinum is 50 troy ounces a sheet or bar. Delivery months are January, April, July and October. Trading may be conducted in other months as determined by the exchange's board of governors. Prices are in dollars and cents per troy ounce. Minimum price fluctuation on platinum is 10¢ per ounce with a maximum daily price fluctuation of $20 per ounce above or below the preceding day's settling price. As with many contracts, there is no maximum for trading during the current delivery month.

PALLADIUM

The NYMEX palladium contract unit is 100 troy ounces weight, with a tolerance permitted of plus or minus 2%. Delivery months for palladium are March, June, September and December. Minimum price fluctuation is 5¢ per ounce, and maximum daily price fluctuation is $6 per ounce above or below the preceding day's settlement price.

Hedging

Hedgers play a large role in the market. Gold product fabricators, in particular, who issue fixed price catalogues without owning gold, generally take a very great risk and over the last few years have generally learned that forward pricing or hedging in the markets is very important.

Gold miners and refiners have also started to use the futures markets more than ever before to hedge forward contracts and production.

The role that the banks play is particularly important to understand, because with their help the bullion professional finances buys from the international physical market, from the

U.S. treasury or from the IMM (International Monetary Market). These, however, are traded in 400-ounce bars, which have to be converted into 100-ounce bars to conform with the other exchanges delivery requirements. (We have given or we will give contract particulars for all precious metals. This information is available to anyone who requires it simply by writing to any of the exchanges.)

Futures Exchanges: How They Work

How an order reaches the floor: A customer who puts in an order would call a registered representative or account executive, or whomever he or she deals with at a bank or brokerage firm. The broker then time clocks the order and hands it to either an order clerk or a main order desk at the particular firm. The order is either telephoned or teletyped to the floor of the exchange.

On the floor of the exchange, the phone clerk working for the firm gives the order to a "runner," who makes sure that the order reaches the clerk or broker at a trading ring who gets paid to execute the order. When the order is executed, a written report (a trade card, actually) goes immediately to the firm's order clerk on the exchange floor. That clerk phones or teletypes the report back "upstairs" to the central order desk from where the clerk received that order. The customer is then notified that a trade has been made.

This circuitous route often results in prices that do not seem to be consistent with the tape of prices sent out of the exchange, particularly in a very fast moving market. Investors frequently wonder why a trade was not executed at a better price. It is difficult for an investor to realize that most transactions really are appropriate. Prices can move so rapidly in a futures market that by the time the execution is finally reported to the customer, many minutes may have elapsed. These minutes can certainly seem more like an eternity in a fast moving market, but they mean that execution prices won't exactly match the tape that an investor is watching.

Regulators from each exchange and the government monitor trading to ensure that markets are well run and maintained. Exchanges hire fairly large surveillance teams. Also, the Commodity Futures Trading Commission has an audits and investigation department. The CFTC was established in 1975 to safeguard customers, promote confidence in the futures markets, and ensure that contracts serve a valid economic function. Many brokers feel that the CFTC is like any government policeman—someone who is there to catch them doing something wrong. I disagree. The CFTC investigates all complaints from customers most of the time at the customer's initiation. It is my belief that a strong CFTC adds confidence to the markets. Safeguarding customers will generally promote public confidence and increase participation in the markets. The CFTC's work is difficult and the terrain hard to patrol. The industry has its share of operators who have abused clients, something that is not hard to do when an investment route is new and offers the potential of high profits.

Unfortunately, some of the greatest wrongs done to the public have come from people the CFTC are unable to pursue. These are "boiler-room" operators who are not members of any exchange or any organized market, and, therefore, do not fall within the jurisdiction of able investigators. Recently several state regulators and the National Futures Association, a new self-regulatory group with some enforcement power, have stepped up their efforts against fraudulent purveyors of precious metals.

The exchange staffs are also very important cogs in the regulatory process. Their day-to-day surveillance, their visibility and their attitude has gone a long way toward insuring that the public is protected at the exchange level. Exchanges have the power to fine and even suspend members and their employees who violate rules aimed at ensuring a free and fair market. They frequently exercise this power.

As previously mentioned, the Commodity Exchange is located at 4 World Trade Center in downtown New York City. It is one of four exchanges that share one trading floor in a very large 22,500 square foot room. In other parts of the room, people are actively trading in other rings—gold, copper, silver and aluminum on Comex, platinum, palladium and gas and oil on the New York Mercantile Exchange (NYMEX); coffee, sugar, cocoa (and soon a Consumer Price Index future) on the Coffee, Sugar and Cocoa Exchange; and cotton, propane and orange juice on the New York Cotton Exchange.

Membership on these four exchanges and seven others in the U.S. (four in Chicago, one more in New York, and one each in Kansas City and Minneapolis) is approved by the membership committee and a board of governors of an exchange. A member can confer corporate privileges on his or her employees. Anyone who wants a membership must buy one from a current member. Membership prices fluctuate. A few years ago a Comex membership was $15,000. Today that membership is worth around $100,000. The NYMEX membership, which was worth around $6,500 in 1974, was trading around $65,000 in 1984.

Floor brokers make up about one-third of the members on most active exchanges. They act on the floor as traders for commodity brokerage firms that deal for customers and so-called "trade house" merchants—commercial firms. Some floor traders believe that being in the ring gives them an advantage in sensing the direction in which a market will move. Floor brokers watch carefully to see how frequently runners are being sent to bring orders into the ring.

Ira Shein was a senior vice president and senior trading member of a world-wide commodities trading firm, ACLI Metal Company. He has been dealing in commodities for over twenty-five years, and in physical metals for more than fifteen years. He is secretary of the New York Mercantile Exchange and a member of its Board of Governors and its administrative committee. He also is chairman of the Metals Committee.

Could you tell me what a terminal market is?

IS: A terminal market is a trade term used for an organized exchange which deals in futures of a particular commodity. In the United States, only three or four exchanges deal in metals. In Europe, the largest one, which is utilized by people dealing from other parts of the world, is the London Metal Exchange. It is organized somewhat differently from the U.S. markets.

What are the elements of a typical trade in the physical market? How does a trade take place and what happens afterwards?

IS: Let's take one of the commodities that are traded on a U.S. terminal market—gold, platinum, silver, copper, palladium or aluminum. Normally the price of a physical metal is related to the price of the futures market by means of a premium or discount, which is generally accepted in the trade. These premiums or discounts may vary from time to time for different reasons, including demand, supply, particular shortages or availabilities, or the passage of time. If, for example, I buy 5,000 ounces of silver...

Who would you buy that from?

IS: I would buy it from a producer who was refining either ore or scrap material.

Would you call up a mine and say to the management of the mining company, I would like to buy 5,000 ounces of physical silver, or would they call you?

IS: Could be either way. There's no protocol in that sense. It depends on the relationship you've got, and it depends on what the particular organization with which you're dealing wants to do. There isn't any single answer to that question. It could as well be from another dealer who already has metal purchased and wishes to sell the metal.

In other words, you could be buying the physical metal from a mining company, another dealer, maybe from an individual or from a foreign government, or from a refinery.

IS: Yes, in fact, it would seem to me that, for the most part, though many mining companies don't sell directly to somebody else, they would sell to the refinery, particularly the smaller ones, and the refinery would in turn be the marketing instrument.

Why would the refinery not hedge on an exchange by themselves? Why would they sell the physical metal to a dealer such as yourself?

IS: The refineries, for the most part, may in fact be hedging. The exchanges are not organized as marketing entities. They're organized as price protection entities and, while any contract on the exchange has delivery as its stated intention, as a practical matter a relatively small percentage of the contracts that are entered into on the exchange are settled by delivery. Also, the contracts which are entered into on the exchange may be for delivery in some subsequent month which is a more liquid month than the spot month, and the refiner may not wish to deliver the material. They may wish to deliver the material now rather than in a subsequent month where they've got their hedge placed. In order to do that, they must enter into a physical contract. Furthermore, there are refineries which produce metal that is not deliverable on the exchange because it doesn't meet the exchange specifications.

That does *not* mean that nonstandard metal is in any way inferior. It simply may be produced by somebody who is too small for the exchanges to recognize or somebody who doesn't produce enough material.

In the case of platinum or palladium, the refiner may be producing material that is not in a deliverable form, even though they may be producing material that meets the standard for purity. They do not produce, for example, hundred-ounce plates of palladium; they may be producing sponge. (Sponge is a form that Russian refineries produce). Palladium sponge is not deliverable on the exchange, and it would have to be converted into plate to make it deliverable. And, for all practical purposes, a substantial amount of consumption in palladium is done as sponge and not as plate, and, therefore, it's a cost factor, and a factor that cannot necessarily be recompensed if the refiner were to convert everything it is producing into the plate which is deliverable on the exchange.

That's interesting, but aren't you also in effect a financier? The reason I say that is that someone may also elect to sell you the physical, because the physical has to be paid for the next day.

IS: In two days.

In two days, but if someone hedges in the futures market on an exchange, he or she will have to wait until first notice day of that particular trading month in order to receive funds from whoever bought the contract. First day notice is the election of delivery for that trading month. Notices of delivery must be made before the future contract becomes spot.

IS: That's correct. Now, of course, one must recognize that all the metals exchanges have spot contracts which they could sell and make immediate delivery, if their metal were in position in the bank and met all of the specifications. So that's only part of the answer. But it is true that, to a degree, a dealer's role involves buying things and paying for them immediately if necessary.

Are the physical markets pretty much the same in the United States and elsewhere in the world?

IS: Trading in precious metals is for the most part an international affair, and the diversity of price between the same commodity in the United States and abroad is usually quite small because arbitrage brings the international markets into close price proximity.

By "arbitrage" you mean the simultaneous purchase and sale of the same commodity in different markets at various prices? For example, buying gold in Chicago and selling in New York at the same time.

IS: Not necessarily the same commodity. It could be price-related commodities, or commodities that could be substituted one for the other, and which, therefore, have prices that tend to move together. But arbitrage does involve the simultaneous purchase and sale in different markets of commodities that are either the same or similar.

And you think that the physical market is a very large market with depth, or do you think that there may be a handful of major dealers that do most of the physical business?

IS: Compared to the futures market, there is considerably less depth and less immediacy in the market because it is not located in one place where all the orders from various places can meet. The physical market is a telephone market where people deal with each other over the telephone, and you can only deal with one person at a time. Furthermore, if you enter into a physical market with any other individual or corporation, that contact must be either delivered or liquidated with the other party only. If you enter into a contract on the exchange, it doesn't make any difference with whom you made the contract; you can liquidate it by taking the other side on the exchange. For example, if you agreed earlier to buy silver, you can liquidate your position by selling another contract in the same market. The physical market has far fewer players because the public does not generally deal in the physical market, is not recognized to deal in the physical market, and would not be accepted as a counterparty by the substantial dealers or others who normally operate in the physical market.

Think of a typical day. Is it conceivable that, in the morning, a South African mining company could sell you some platinum for physical delivery, and in the afternoon an American automobile company could buy some platinum from you also for physical delivery for production, and that all of this could take place within the same day at various prices?

IS: Yes, that is a scenario that conceivably could happen. Of course, that's where the futures markets come in, because if I were to buy in the morning as you postulated, I would either have to take a risk or I would have to hedge my purchase, and normally a dealer operates to some degree on risk. That's a matter of judgment and a matter of choice—but, substantially, a large part of the metal or any commodity that is purchased would be hedged on the exchange. In other words, I would sell an equivalent quantity of what I bought from the producer on the exchange, at a differential which I considered satisfactory for my purposes of price protection.

In your example, in the afternoon if an automobile company wished to buy from me, I would sell to them and in turn buy back on the exchange. In the interim, I would have protected myself against any untoward fluctuations that could have cost me aggravated losses.

Obviously, I may have reduced my profit, but the exchange fulfilled its insurance function by protecting me against substantial losses. Basically, even if the price has not moved in the futures market, my profit would be made because the discount at which I would have purchased the metal from the mining company would be less than whatever discount I might have to give to the automobile company. In other words, it is the difference in the premium and discount from which one can make money, even if the price on the futures market remains the same.

Do you keep a book with buyers' and sellers' names readily handy? Is that how you normally operate in the morning? So that you know whom to approach if you buy something or if you sell something? Or do you wait for someone else who is a physical dealer to contact you?

IS: Both. I have the names of the people with whom I am dealing consistently, and I am on the phone with them. I may call them or they may call me, and it totally depends upon the circumstance. There is no uniform answer to that question, since both happen.

Is it essential for you to have the availability of a news ticker for the financial markets during your day-to-day operation?

IS: It's generally helpful to the extent that news stories of important events are price-influencing things. Therefore, at any given time, I must be aware of what is influencing the price to make a determination as to whether I want to make a particular move.

Is there a place where physical prices are quoted on a daily basis other than in the spot prices of a spot month on one of the exchanges?

IS: You, can get them from some of the markets. Some of the newspapers do print these. There are quotes that are made at an instant in time and therefore have immediacy only for that time but serve as benchmarks for trading. For example, the Englehard silver price comes out each day at 11 A.M. (New York time), and the Handy & Harmon price at noontime. There is the London bullion fix for gold in the morning, and, in the afternoon, there are other prices that are published from time to time. Weekly, one could get prices in various publications such as *Metalsweek*, which will give you a bid and offer as well as whatever producer price is established by the producers. So there are various publications, but none of them are, of course, valid for other than the time at which they've stated. I should disagree for a moment and point out that there are times when somebody will say, "You're too expensive, *Metalsweek* says it's worth this or that," and the answer, of course, is, "Go buy it from *Metalsweek*."

Is it really important to keep abreast of the physical market if you plan to trade the futures market?

IS: It is very helpful. For all practical purposes, in the precious metals the price-making influences in most cases are not really the demand-supply of the commodity, but the demand-supply of contracts. We have rarely faced a real shortage of gold or a shortage of silver. At a price, they were always available. There has at times been a shortage of palladium for the simple reason that the production of palladium is much more concentrated in the hands of a relatively few producers. For this reason, the price of prompt delivery of palladium does not reflect storage and interest expenses. But, generally speaking, the price-making influences are the demand-supply of contracts as influenced by the news factors and the idiosyncratic thinking that goes into people's opinions as to whether prices should go up or go down or whether they want to buy or sell. So, that is a partial answer to your question.

Is that the dog that wags the tail or the tail that wags the dog?

IS: Before I answer that question, this has not always been the case in copper, although to a greater or lesser degree it's also true that there have been times when substantial strikes in the copper industry or transportation bottlenecks or other things of that kind have created real demand-supply situations as distinguished from what you might find in gold or silver.

As to which is the dog and which is the tail question, the role changes from time to time. The futures market sometimes makes the price; rarely does the physical market make the future price. However, one must distinguish such a factor as, for example, the London gold fixings, which may create the situation for the opening for the markets in New York, because they are such widely disseminated prices.

Terry Martell is senior vice-president of economics marketing for Comex.

What kind of work does your department do?

TM: I oversee the operation of several departments within the exchange, including research and marketing. Our research area devotes its attention to new product development. The marketing area develops support materials, conducts public relations activities, coordinates the exchange's advertising effort, and undertakes many other projects to promote our markets.

Where can a new investor get the information he or she needs?

TM: Each day, the volume and open interest in each contract market are published in *The New York Times*, *The Wall Street Journal*, and various other trade publications. Sometimes this is all the information that is needed. However, those individuals who need more information and who are concerned, for example, about warehouse stocks or the pattern of open interest should obtain a copy of Comex's *Daily Market Report*. In addition to the volume and open interest statistics, the report provides information on stopped notices (who is accepting or delivering gold, silver, copper, and aluminum), how much of the various commodities is stored in Comex-licensed warehouses throughout the country, and additional statistical information that might be useful to individuals who are trying to form an opinion on the market.

Does the exchange have occasional lecture series or programs, or would one have to go through a member firm to get involved with that sort of thing?

TM: From time to time, Comex conducts seminars and workshops on particular contracts traded at the exchange. Whenever a new market is introduced at Comex, an extensive training program is held. The exchange offers brochures, audio-visual materials, and other materials which might assist someone who is trying to gain an understanding of the markets.

Member firms might hold seminars, but one would have to contact the firms directly to find out when such programs are offered.

What type of research would you suggest for a new investor, outside of brokerage house or trade house recommendations?

TM: If I were dealing with the metals area, I would ideally want to know something about the underlying cash market. By the cash market I mean the actual commodity being traded in the physical market. In the case of silver, it would be silver bullion. I would want to know something about who uses silver, how much silver is produced, and how much silver is consumed. I would also try to identify the kinds of political and economic events which affect the price of silver. More generally, if there is going to be an increase in overall economic activity, one can assume that there is going to be an increased demand for silver. The increased demand often leads to rising prices. Moreover, in addition to researching the cash market, the smart investor also knows the different investment vehicles that are at his disposal. For example, I think that in the metals area, purchasing options might be a good way to start acquiring investment experience.

Basically, you're saying that an option which gives the buyer the right to buy or sell a specified commodity at a specified price on or before a specified date, limits the investor's exposure? In other words, the amount of premium that you're going to pay is all the money you can possibly lose on a transaction?

TM: That's right. So you can pick your shots, so to speak. If you think gold or silver is going up, buy a call, pay the premium and sit back and watch. See if your expectation turns into reality without the need to be concerned about day-to-day volatility.

Do you have any suggestions or thoughts on the best kind of research for predicting a market?

TM: Some people use their computers to follow trends and use those trends to make assessments about future price moves.

Do you favor these technical advisers over fundamentalists who study market conditions?

TM: That's a good question. My continuing problem with technical research is that it assumes patterns from the past will repeat in the future. That's a difficult assumption to buy. I never felt comfortable with someone else's technical trading technique. I always felt that if the trading was so successful, why is the guy selling his technique and not trading the market. I approach these technical trading strategies with a good bit of caution. I don't mean to suggest that this analysis be discarded entirely, because we all know that certain support levels and prices are on people's minds.

For example, assume gold is selling at $300. Based upon the charts, people may think that there will be significant additional selling if the price breaks $280. The additional selling triggered by the price's breaking a support level may result in another drop in price. A trader should be aware that certain kinds of price patterns are likely to trigger buy or sell orders. A small trader should know those things and defend against them. But he or she also should be cautious about spending a lot of money for someone else's technical trading advice.

Fundamental research has to play an important part in being continually successful, doesn't it?

TM: I believe that. You have to know the fundamentals. You have to know the supply and demand situation, and have some sense of who's in the markets and why. Again, I don't want to always bring the conversation back to options, but these are some of the reasons why, particularly for the new investor, I think options are an excellent investment vehicle. An option limits your risk and also gives you staying power.

FIVE

Options—Alternatives for the Uncertain Investor

An interesting new product known as options is now actively traded on many commodity exchanges. Options are rights to buy or sell commodities, or stock on stock exchanges. The best way to describe an option is to think of it as "rent." You pay a potential seller, or landlord, a premium that gives you the right to buy or sell a contract at a certain price within a set period of time. It is important to note that in purchasing an options contract, you acquire the right but not the obligation to purchase the futures contract. If the option expires without your exercising your right to buy or sell, all you lose is the premium, the "rent" you paid to the option seller.

There are two basic kinds of options—one giving you the right to buy a contract and one to sell a contract. A right to buy is a "call option," because you *call* it away from the owner. The right to sell is a "put option," because you are *putting* the contract into someone else's hands.

The advantage of buying an option is that it gives you the ability to command a position in the market for a minimal amount of dollars. You have a clearly defined risk in that you can only lose the amount of money that you have put up, the premium. (The seller of an option—also called the "writer" of the option—has a far greater risk because he or she can have a profitable contract taken away. Most speculators are buyers of options, not sellers.)

Options are complex and not widely understood because they are relatively new. Many account executives are not well trained in their use.

Let's concentrate on the gold options, traded on Comex. Remember that as a buyer you have the right but not the duty to purchase or sell the gold, the underlying futures contract. An option provides you with a defined degree of risk. The extent of your risk is neither more nor less than the cost of your option. This means that you can never lose more money than you have paid for the premium.

Everyone has a different conception as to what a premium should be, and that is why you have buyers and sellers. Some components of premiums are:

(A) Interest rates (carrying charges).
(B) Volatility of the underlying futures contract (risk factor).
(C) Time. (The longer the time, the higher the premium. As an option nears expiration, if the underlying commodity becomes less valuable or slips in price, the option value will decrease proportionately.

As an example, say that you have purchased in February a call to buy 100 ounces of April gold at $400 per ounce for a premium of $5000. Gold is currently trading at $375 per ounce. This means that for $500 you have the right to buy the gold contract from the option seller. The exchange ensures your transaction. If gold goes to $406 between purchase time and April, you would certainly exercise your right to "call" the contract.

At what point will you have a profit? Your costs are $500 for the call and $40,000 ($400 per ounce times 100) for the 100-ounce contract. If your exercize your right, your real cost is $40,500 per contract. If gold goes to $400 an ounce, do you have a profit? No! Because you purchased the call for $500. You only have a profit if you can receive more than $40,500 for your 100 ounces of gold.

Now, let's look at puts. As we said, a put means that you have the right to sell a gold contract. Why would you want to purchase a put? A seasoned trader might do it for "insurance" against a decline in the price of the contract. Here's how it might work.

Say that gold is trading at $400 an ounce. You purchase an April $400 put, for a premium of $500. If gold goes below $400, you could buy a 100-ounce contract at the current market price, say $375, and exercise your put. You would make the seller of the option buy 100 ounces of gold at $400 despite the going price of $375.

In another situation, suppose you are long (have already bought) gold futures. Your gold cost you $400 an ounce. But you feel queasy about maintaining the position because prices are beginning to fall. You could just sell out your futures position, but over the long term prices will go much higher. This is the perfect opportunity to purchase a put option. If gold goes down, you will simply exercise your put option and deliver your long position to the seller of the put option. If gold rises, on the other hand, your put option will expire unexercised, leaving you minus the premium price. But with any luck you can more than recover your loss by selling your gold future at a higher price than you paid for it.

Now suppose you have decided that the price of gold is going to drop sharply. What can you do? Again, you could purchase a put. But you also can sell a call (give someone the right to buy your contract) without owning gold. "I am happy to receive a premium for selling gold in the future," you would think, because you do not believe that gold will go up and that the call buyer will exercise the option. In other words, if you are bearish, buy a put, or sell a call; the only thing *not* to do with a bearish outlook is to make an outright buy of a call or to sell a put (giving someone the right to make you buy). The outright buyer of a call (outright meaning buying without any other position to protect yourself) generally feels the price will go up. Someone who already has a short position (a contract he or she does not own in the hopes of repurchasing it later at a lower price) might also want to purchase a call option as insurance against prices suddenly rising. By buying a call option, an investor can limit the risk of a sudden move in prices.

In summary, there are six major positions in the options market to match your view of price movement. A bullish outlook, the expectation that a price will go up, calls for purchasing a call, selling a put, or simultaneously buying the call and selling the put. A bearish view, of course, leads to purchase of a put, sale of a call or simultaneous purchase and sale of both. The simultaneous transactions, of course, are much more risky—meaning they also hold out the potential of greater pain.

Any time one sells an option, one receives a premium, which, of course, is income. For that income, one has a tremendous amount of risk. Suppose you have sold a put. You are giving someone else the right to sell you 100 ounces of gold at a fixed price. If the price of gold were to drop sharply, that put would be exercised against you, and you would have to buy the gold at a price substantially higher than the market. Such risk may be unwarranted by the amount of the premium you would receive for selling the put. Similarly, if you sold a call, giving someone the right to buy gold from you, you would be at risk in the event of any upside movement in gold. A general principle of options is that writing—that is, selling—is much riskier than buying. Your loss is limited to the premium if you buy; it is unlimited if you sell.

If you have purchased a futures contract and sold a call, it is known as covered writing. But even though a covered write guarantees you premium income, any upside move in the price of gold will limit your profit. You have given away this profit potential by selling the option to the buyer of the call.

Another aspect of options trading is the relation of option pricing to the date of expiration. If you buy a call option today and the price of gold moves up, there is a very good chance that the option you hae bought will increase in price. You could turn around and sell your call at a profit. If you buy a put option and the day after your transaction the price of gold goes down, there is a great likelihood thaathe put option you have purchased will become more valuable. As you now see, options can be traded just as futures contracts can be traded. There are almost always others who will buy or sell the option according to the price behavior of the underlying futures contract. Remember, then, that you are trading the premium and not the underlying bullion. This helps to make options trading more complicated than futures contract trading alone. With options one should always consult with a knowledgeable individual at every step of the way.

Case Study

Selling a call option: Let us say that you are a long holder of a futures contract in gold. Assume that your gold has an average cost price of $375 an ounce. Gold is now trading at $385 and you do not expect the price to go much higher in the near future. In fact, you are concerned that gold may soon drop back to $375 or maybe $350, but for a variety of reasons you want to hold your underlying futures contract. To provide additional income for holding your investment, you could sell one 100-ounce gold call option. It is February, and you can obtain $5 per ounce for a $385 call that expires in April. this means that, to all intents and purposes, you have sold your gold out at $390. The $500 premium goes into your exchange account as a credit on your futures position. There is no loss in your position even though you paid $375 for your gold and the current market is $385. But if gold dèclines below your purchase price by more than $500, you begin to realize a loss. In this case, you can place a $375 stop loss on your long position. This means that your gold position will be sold automatically if gold goes below $375.

One fine Friday afternoon, gold reaches $374 and your broker sells out your long position. You feel very comfortable because, after all, you have escaped a little better than even. You paid $375 for the gold, you got out at $374 with a $100 loss, but by selling the call option you have received a $500 premium. Without commissions and other charges (such as futures fees), you have still netted $400 on your position. (Because commissions aa so highly negotiable and subject to frequent change, I cannot recommend a commission level or factor in a commission price onany trade.)

That was Friday afternoon. Now it is Monday morning, and some political news hits the market which causes gold to open $16 higher. This makes you wary because you have sold a call, and you are obligated to deliver to the buyer at the striking price of $385. As volatility increases, premiums turn up sharply. At $390, the buyer will certainly exercise the right to purchase the gold from you at $385. Even though the purchaser is buying at $385, five dollars below the market price, he or she has recaptured the $500 premium paid.

You repurchase at the current price of $390, incurring a trading loss of $16 per ounce, from $374 to $390, or a $1600 loss for the 100-ounce contract. Your premium income was $500, so therefore you have $1,100 plus two commissions as an outright loss. This is one of the pitfalls of selling an option.

Dealer Options

Besides the exchanges, four private firms have government approval to issue precious metals options, the largest of these being Mocatta Metals Corp. It is noteworthy that dealer options have wide bids and offers, are thinly traded and have involved many firms in litigation.

A dealer may have a position in physical bullion with a current market value of $38,000. The dealer now sells an option with a striking price of $400 an ounce against a futures contract valued at $380 per ounce for delivery two months later. The dealer receives a premium of $400, and gets interest income on it. Unless gold rises to over $400 an ounce, the underlying futures contract will not be called by the buyer.

When dealers want to take advantage of a probable rise in interest rates, which can cause a decrease in the price of gold, they may borrow warrants, or certificates of ownership, from lenders at a given rate. (The lenders may be other dealers, or foreign or domestic banks that are passive holders of bullion.) Dealers will then sell futures contracts on an exchange, creating a certain balance in their own account. These credit balances are secured by good-faith deposits (called margins), which are posted in Treasury bills by the dealer, and also provide interest income.

When the price of gold falls, the dealers may repurchase the gold they sold at a cheaper price. These operations are normally short-term and provide liquidity in the markets. This liquidity encourages retail investors, because they know someone will be there to buy and sell their contracts.

Dealers expect that commission houses, which transact for investors, will normally sell futures contracts that are close to expiration and will simultaneously purchase another futures contract month. The reason for this is simple. As the futures contract becomes spot, and subject to actual delivery, the commission houses request clients to immediately post the full price of the contract. Most retail clients do not want actual physical delivery of the commodities, but do wish to continue owning the futures contract by maintaining the small margins which exchanges and brokers mandate.

The difference can be one of paying an additional $35,000 cash for a deliverable position as against a $3,000 margin on a futures contract.

The following are various investment situations involving options.*

*By permission of the Commodity Exchange, Inc. (COMEX).

BEARISH TRADING STRATEGIES

Opinion: Strongly bearish
Strategy: Buy put options
Risk: Limited to the cost of options
Potential Profit: Unlimited—Intrinsic value of the options at expiration less initial premium paid

Example
An investor anticipating sharply lower gold prices buys a $500 put option on April gold futures for a premium of $24 an ounce or $2,400. Should April gold futures decline from $496 to $400 an ounce, the investor will earn a profit of $7,600 ($10,000-$2,400). However, should the April $500 put option expire with April gold futures trading above $500 an ounce, the most the investor can lose is his initial $2,400 investment.

Strategy: Buy Put Options

Futures Price At Expiration	Option Premium At Expiration	Gain (Loss)
$400	$10,000	$7,600
$450	$ 5,000	$2,600
$480	$ 2,000	($ 400)
$500	—	($2,400)
$550	—	($2,400)

Comment
Purchasing put options allow a bearish investor to earn highly leveraged profits with a limited, known risk. An investor who is extremely bearish might choose to buy out-of-the-money options, which can offer the greatest returns per dollar invested in a major price decline. An investor who is only mildly bearish might purchase in-the-money options, or choose to utilize other strategies detailed in the following pages.

Opinion: Strongly bearish
Strategy: Sell futures and buy call options
Risk: Limited to time value of the option plus the amount, if any, that the option is out-of-the-money
Potential Profit: Unlimited—Futures profit less option premium paid

Example
An investor anticipating lower gold prices sells the April futures contract at $496 an ounce while simultaneously protecting his position by purchasing an April $500 call option for a premium of $21 an ounce, or $2,100. Should futures prices fall to $400 an ounce, the investor earns $9,600 on the futures position but forfeits his $2,100 option premium, for a net profit of $7,500. If futures prices rise to $550 an ounce instead, the investor's $5,400 futures loss is cut to $2,500 due to a $2,900 profit on the call option.

Strategy: Sell Futures, Buy Call

Futures Price At Expiration	Futures Gain (Loss)	Option Gain (Loss)	Net Result: Gain (Loss)
$400	$9,600	($2,100)	$7,500
$450	$4,600	($2,100)	$2,500
$480	$1,600	($2,100)	($ 500)
$500	($ 400)	($2,100)	($2,500)
$550	($5,400)	$2,900	($2,500)

Comment
The short futures, long call position is often called a synthetic long put option because of its risk/reward characteristics. Thus, the results for the transaction at various final futures prices are nearly identical to the results in the previous long put example. Like the synthetic long call, the synthetic put is highly flexible, and provides a continuous flow of cash into and out of the investor's account due to the daily mark-to-market of the futures contract

Opinion: Mildly bearish
Strategy: Bear put spread (vertical): Short the lower strike put, long the higher strike put
Risk: Limited to the net premium paid (occurs when both options expire out-of-the-money)
Potential Profit: Limited to the difference between the strike prices of the two options less the net premium paid (occurs when both options expire in-the-money)

Example
An investor who is looking for a modest price decline, but not a major price collapse, sells an April $480 put for a premium of $15 an ounce, while buying an April $500 put for a premium of $24 an ounce. The net premium paid on the transaction is $9 an ounce, or $900. If the options expire with April futures trading below $480 an ounce, the investor will realize a profit of $1,100 ($2,000 difference between the strikes less $900 premium). Should April futures rally instead, and close above $500 an ounce, the investor will lose the $900 that he paid for the spread.

Strategy: Bear Put Spread

Futures Price At Expiration	Gain (Loss) on $480 Put	Gain (Loss) on $500 Put	Net Result: Gain (Loss)
$460	($ 500)	$1,600	$1,100
$480	$1,500	($ 400	$1,100
$500	$1,500	($2,400)	($ 900)
$520	$1,500	($2,400)	($ 900)

Comment
The investor who uses the bear put spread in a down market rather than simply buying a put gives up the benefits of a major price decline in order to increase his gain from a moderate price decrease. The choice of a spread effectively caps the transaction's potential profit at $1,100 when the price of gold reaches $480 an ounce. Nonetheless, had the investor simply bought the $500 put, rather than utilizing the spread, he would have lost $400 on a gold price decline to $480 an ounce. For about the same dollar investment, the investor could have utilized two bear put spreads and earned a profit of $2,200. On a major price decline, however, the investor would almost always do better with an outright purchase of a put.

Opinion: Mildly bearish
Strategy: Bear call spread (Vertical): Long the higher strike call, short the lower strike call
Risk: Limited to the difference between the option strike prices less the initial premium received (occurs when both options expire in-the-money)
Potential Profit: Limited to the initial premium received (occurs when both options expire out-of-the-money)

Example
A mildly bearish investor sells an April $480 call for a premium of $3,080 ($30.80 an ounce), while at the same time purchasing an April $500 call for a premium of $2,100 ($21 an ounce). The net premium received on the transaction is $980 ($3,080 – $2,100). If the options expire with April futurbs trading below $480 an ounce, the investor will make a $980 profit. Should April futures rally instead from $496 to $520 an ounce, the investor will lose the $20 difference between the strike prices, or $2,000, less the $980 premium received or $1,020.

Strategy: Bear Call Spread

Futures Price At Expiration	Gain (Loss) on $480 Call	Gain (Loss) on $500 Call	Net Result: Gain (Loss)
$460	$3,080	($2,100)	$ 980
$480	$3,080	($2,100)	($1,020)
$520	($ 920)	($ 100)	($1,020)

Comment
The bear put and call spreads both reflect a mildly bearish opinion about the market. Like the put spread, the call spread increases the benefit to the investor from a small price move, but eliminates any additional benefit that would accompany a major price decline. Unlike the bear put spread, however, the bear call spread is a credit transaction (i.e., net premium collect) and thus requires margin.

Opinion: Mildly bearish to neutral
Strategy: Sell call options
Risk: Unlimited
Potential Profit: Limited to the premium received

Example

An investor who is neutral to slightly bearish on the gold market sells an April $500 call for a premium of $2,100. If the option expires with April gold trading below $500 an ounce, the investor will earn the entire $2,100 premium. However, should April gold rally from $496 an ounce, the investor faces potentially unlimited losses. For example, if the option expires with April futures trading at $550 an ounce, the investor faces a loss of $2,900.

Strategy: Sell Call Options

April Futures at Option Expiration	Gain (Loss) on $500 Call
$470	$2,100
$500	$2,100
$520	$ 100
$550	($2,900)

Comment

The call option writer will earn the time value as income in a sideways to slightly lower market. However, his income is limited to the extent of the option premium. Like the spread trader, he gives up any additional benefits which would accompany a major price decline to increase his percentage return in a trendless market. Writing in-the-money calls offers the greatest potential returns if prices decline significantly. However, it also involves greater risk in the event of a market rally. Writing options involves unlimited risk, and thus requires margin.

Opinion: Mildly bearish to neutral
Strategy: Sell futures and sell put options
Risk: Unlimited
Potential Profit: Limited to time premium received, plus the amount, if any, that the option is out-of-the-money

Example
An investor who believes gold prices will be steady to slightly lower over the next several months might sell April futures at $496 an ounce while selling an April $500 put option for a premium of $2,400. If futures prices fall to $450 an ounce, the investor will earn $4,600 on the short futures position while absorbing a $2,600 loss ($5,000 – $2,400) on the short put option, for a net profit on the transaction of $2,000. If futures prices were to rise instead to $550 an ounce, the investor collects the $2,400 option premium, but loses $5,400 on the futures position, for a net loss of $3,000.

Strategy: Sell Futures and Sell Put

Futures Price At Expiration	Futures Gain (Loss)	Options Gain (Loss)	Gain (Loss)
$480	$1,600	$ 400	$2,000
$500	($ 400)	$2,400	$2,000
$530	($3,400)	$2,400	($1,000)
$550	($5,400)	$2,400	($3,000)

Comment
The short futures, short put position is another type of synthetic transation, in this case, a synthetic short call. Not surprisingly, the dollar results for this transaction are nearly identical to those in the previous short call example. However, the synthetic position allows for greater market flexibility.

Gary Glass has been a trader and member of the Commodity Exchange for over ten years. Prior to that he was in charge of Godnick & Sons, a securities and options dealer. He began his career on Wall Street at Filer Schmidt & Co. He is one of the pioneers in the precious metals options market.

Do you trade your options on a daily basis?

GG: Yes.

Is there anything you do to prepare yourself in the morning for the day's trading? What do you read, before you enter into an options transaction for the day?

GG: I must say I go back one step to the previous day. That evening I input all the prices from the day before into my computer. I have an IBM PC, and the program I use was made for me by a head programmer at Columbia University. I input the highs, lows and closes, and the close the underlying gold, and then use my own calculations and run the numbers. I cross-check them with an interest function, by reading through that evening's rèpo rate, which I get from the telerate machine.

That means the interest rate, right?

GG: Or I take a T-bill rate, the current yield of a 90-day bill. That is my preparation in the evening. Each morning I read both The *Wall Street Journal* and *The New York Times* and I cross-check my prices with the closing quotes that are posted in the paper. I study them because I see a lot of spread differentials that may be out of line.

What are spread differentials?

GG: Spread differentials are the differences between strike prices in an options series. For example, August strike prices for gold ranged between 360 and 530, meaning $360.00 anaunce and $530.00 an ounce. August gold was trading somewhat under $400 for the futures. The $400 strike is the most accurate, but I will offer to trade puts and calls on August contracts from 360 to 530. What I mean by that is that I might buy a 360 call and simultaneously try to sell a 400 put. The difference between what I pay for a 360 call and what I get for a 400 put would be a nice break, because a 400 strike price would be a more expensive option than a 360 strike price, which is far from current value. That would be a bullish spread, particularly if I bought a 360 call and sold at 400. When you're bullish, what you want to do is buy a call and sell a put, because you wouldn't mind buying the underlying futures contract and receiving a premium for it. The premium which you receive for your put could be as much as the dollar amount that you pay for the call.

Now let me ask you about the pricing of the options. Do the premium prices react very quickly in response to underlying futures prices?

GG: Yes, as volatility gets lower, there is less risk in owning an option. The most important factor in pricing options is volatility. Second in importance are interest rates.

If the option has longevity, the closer the option strike price is to the traded futures market current price, and the difference of the greater actual dollars can be calculated as the premium. If the option does not have longevity, the dollar amount involved will be less.

When you buy an option you are buying time and paying an interest rate to boot. On the other hand, your total risk is only your total premium paid.

There are some sophisticated techniques used between options and futures primarily by institutions to create an interest rate return. Can you explain how this is done?

GG: There is basically one technique with several variations. It is called a conversion. That means simultaneously acquiring a put, the right to sell a futures contract and a futures contract, and at the same time selling a call option. So by selling a call option you're giving someone else the right to buy the futures contract that you have acquired. But the put option gives you the right to resell that futures contract to someone else. It is a neutral position.

How do you make money with a neutral position?

GG: The tendency is for the market to become inefficient sometime during the day, so that occasionally the future option will trade at less than its actual value in relationship to the call and the futures market. Or conversely, the call will be undervalued against the put and the futures market. The conversion becomes a basically riskless position. The potential risk in this transaction occurs only if by chance the options should expire with the future at the exact striking price.

Do you recommend these strategies for retail investors?

GG: I don't think the general public could spend the time or save enough money on all the transactions. As floor traders, we get a much lower commission rate than the general public and are, therefore, in a position to trade quickly on small moves.

I've heard that many option trades made during the day are done by professionals who are using computer techniques. Would you say that most people entering into options are using some sort of computerized option technique, either for market or interest purposes?

GG: Every market maker I know uses some form of profile or program.

Would you say the number of floor traders, those who trade for their own account, has increased significantly over the past year or two?

GG: Without question. First of all, as far as I am concerned, they are the future of this business. There is probably no better tool, in my opinion, particularly in a commodity business, whereby you can predetermine exactly what your risks are.

How important is it for precious metals traders to keep abreast of options, whether they are using them or not?

GG: Very. In fact, at this point I know of two major metals firms that would probably first quote the option market before even attempting to assess the metals range. Comex may do more volume per capita on options than the underlying futures contract, though it might never even be visible. So, in fact, if a metals dealer were to use the market, he or she might be able to effect trades without other competitors knowing what is being done.

How useful are options for someone who wants to speculate on a long shot, on a very distant striking price far above the current market? It wouldn't cost much in premiums.

GG: There have been a number of occasions in which I have done that. On one book of mine I took a substantial loss approximately a year ago. February gold was trading at about $522. A major metal house came in a week or two later and bought 1,400 puts. The option it bought was approximately $122 away from the current market. Fortunately for the dealer and unfortunately for me, the market dropped $130 within approximately a week and a half. And those puts that I sold for a dime, meaning $10 per contract, cost me $1,900 to buy back, so whoever had taken the other side of that transaction had a phenomenal capital appreciation.

Do you still recommend using options? And if so, do you have any special suggestion for someone entering the markets?

GG: Options are a tremendous tool. There is a good book that I'd recommend, but it's for professionals such as people on the trading floor. It was written by Lawrence G. McMillan and it is called *Options As a Strategic Investment*, published by The New York Institute of Finance. It is probably the most comprehensive book that I have read on options. However, I woularecommend that the beginner read all the exchange publications prior to tackling option trading.

You spend all of your day trading in the ring, making markets, buying and selling. Is there anything that you do besides preparing yourself mentally? Is there anything that you physically do so that you don't lose your stamina during a hectic day?

GG: After doing this ten to twelve years, I can say that the most important thing is having a good pair of crepe shoes.

SIX

Opening a Brokerage Account—How to Select a Firm and Maintain an Account

It is important to find the right broker, bank or physical metals dealer to work with you. A dealer in a national contract market falls under the jurisdiction of some kind of government regulator, and is subject to reporting requirements. Most commodity exchanges will be glad to furnish anyone with a list of all the member firms which are members of that exchange. Anyone wanting to participate in transactions should begin by writing to a commodity exchange research and information department and ask for this list.

Once this list is obtained, try to find a well-established, reputable, long-time exchange member firm to service your account. Depending on your needs, you must further decide whether you would like a full-service broker, a discount house, a broker who specializes in precious metals or a physical trading house. Certainly the requirements for clients will vary from firm to firm. Some require that a member of the public have a net worth of at least $50,000 plus substantial deposits of good faith, or margin. I have always found it most expedient to advise clients that, when depositing margin with a broker, they deposit U.S. Treasury Bills, which are fully acceptable for margin, instead of cash. The reason for this is that while you are engaged in a commodity operation, there is no reason for your good-faith money not to earn you the current interest rate as well. Commission requirements are posted at most of the major firms. Do not look for the lowest commission, because the firm with the lowest commission may be the most expensive route in the long run. Commission rates, however, are highly negotiable. But I have found that the more one is willing to pay in commissions the more service one will receive. That service can sometimes be so valuable as to offset all commission considerations.

If you plan to diversify your precious metals holdings, it would be important for you to deal with a full-service broker. This means that the precious metals holdings of any portfolio must not be totally in the futures markets. Some can be in futures markets, but others can be in the physical markets and/or in securities related to the precious metals markets.

Donald Silver is president of the Maine Farmer's Exchange (MFX Commodities), a clearing member of the New York Mercantile Exchange. He is a veteran of the markets and has developed his own computerized clearing facility.

As president of a firm that is a clearing member of NYMEX, how do you explain margin to customers?

DS: Margin money is a deposit of good faith. It serves as a guarantee that a customer will make good on his transactions, and also protects him or her by letting him know on a daily basis how much he or she has to add (or has in surplus) in the account.

A customer has to deposit an initial margin account with us to open a speculative account. Exchanges require different margins on different contracts, and initial margin varies among clearing members. Most margin variations in new accounts depend on the type of commodity, the contract size and the estimated trading positions that a customer will take. It is important to point out that we are required to keep any surplus margin money in a completely separate, or segregated, account in the bank. In case of a firm's collapse, a customer's money can thus be accounted for and extradited.

What is minimal margin?

DS: The figure set by an exchange which a clearing member has to deposit with the exchange's clearing house. It varies as contract prices fluctuate.

Is it normal practice for a clearing member to ask a customer for margin over and above the minimum required by the exchange?

DS: Yes, we ask for a deposit generally one limit move above minimum. (A limit move is the highest and lowest price an exchange will allow a contract to hit during one day's trading.) We tell our customers, based on many years of experience, and losses, that they can be in trouble if there is insufficient margin and the market goes the limit.

What is pyramiding?

DS: Its an objectionable practice, which we discourage. When someone finds that a contract owned increases in value, he or she uses the increased equity in the account to make additional purchases without really putting up new cash. It's dangerous, because prices can drop as quickly as they rise. A surplus can quickly become a deficit. Then, instead of having a margin call for one contract, you have a margin call for, say, three contracts. And your equity can disappear very fast when you triple a margin call from one to three. This is where an account executive must guide a customer and not let the customer rule. This is particularly true when dealing with large traders. I think you have seen problems associated with pyramiding in the silver market.

Once a customer opens an account with you, what happens when he or she makes a trade on an exchange?

DS: The money is transmitted from the firm to the exchange's clearing house to meet its margin requirement. (A clearing house does accounting for an exchange. It matches buy and sell orders, and gives each clearing member a record of its account on a daily basis.) The margin money stays with the clearing arm until the contract is liquidated (when the contract expires or is offset by taking an opposite trading position).

If the value of a purchased contract increases, the customer will have equity added to the account. Should it decrease in value, there will be a shortage of funds in the account, and a margin call will probably be made. The customer will be asked for additional deposits.

As a clearing member how quickly do you have to put up variation margin—based on the changes in daily settlement prices—with an exchange?

DS: By noon of the following day. That is why New York exchanges require that a member have a representative or an office in New York (and in Chicago for Chicago exchanges). Margin over $100,000 is paid with a certified check; under $100,000, you can send an uncertified check.

How do you get money from the exchange after you've liquidated your position, and the value of your position has increased?

DS: Based upon liquidation, you get a check from the exchange on a daily basis.

So a clearing member receives checks on a daily basis and gives the exchange checks on a daily basis? In other words, the exchange's clearing house evaluates all your positions every day, and then either sends you money or gathers money from you?

DS: Absolutely. It is then up to the clearing member, depending on the different customers' positions, to assure that everyone is policed on computer to see if margin calls should be issued. We calculate all positions, equity runs and margin requirements for customers at the close of the exchange every day. It is run on a computer sheet that shows equity of each customer as an individual, whether it be margin call required or an equity increase.

Do you feel that some large traders try to dictate their own margin terms to firms?

DS: Absolutely. And it is a danger to let any customer dictate and run your business. Rules are rules and they are established to protect the exchange and the clearing member, and should be enforced. We have learned through many years of experience and losses that if we don't have the margin and you go to limit moves, then we're both in trouble. For us, it's bad business. It takes away from the working capital. No matter how good a customer is, if he or she is losing money, you will have an unhappy customer who may be slow to comply.

Besides margin, what do you require from a client about to open up an account?

DS: We must receive a financial information summary and see that the individual meets net worth requirements. If we don't know the individual, we check out recommendations. One of the most important forms a customer must complete, which is required by the CFTC, is a risk disclosure statement. In addition to having it signed by a new client, we must verbally explain it to anyone who has never traded commodities. If we do not fulfill this obligation, our company may be liable for any losses sustained by the customer. Before opening any account, all papers must be completed, the interview and the risk disclosure statement signed, and margin must be in our hand.

Let's return to what happens to your account at the exchange, and see how it affects customers. We know you get a daily accounting of your positions. After you have verified your position with the exchange, how do you collect funds from the customer who has the position? How does the customer usually send you the required funds?

DS: By wire. We used to accept checks from the West Coast, but now we require that all money be wired the following morning to our bank.

So every clearing member maintains an account with a commercial bank, and the commercial bank can receive funds for the clearing member?

DS: That's correct. And we are very strict about collecting margin. It is probably one of the most important responsibilities that an account executive can have to the executive's own clearing firm or employer.

Erroneous trades can be made on the floor of an exchange. How quickly can you discover that you have been given a position by mistake?

DS: On the following morning when we go to check out. The exchange has to agree with our computer run at the end of the day. Your first indication of error would be a money differential between what the exchange is saying you owe or should receive and what our records show. It indicates an error either on our books or on the exchange books, such as applying someone else's trade to our books. An error can be rectified by the following morning after we reconcile the exchange's clearing sheets with our sheets.

Manfred Rechtschaffen is senior vice-president of investments at Prudential-Bache Securities. He has been with the firm for over twenty years and specializes in private and institutional commodities accounts.

Can you define the two types of accounts opened with a broker—speculative accounts and hedging accounts?

MR: A hedging account is used in a situation where the client's trades are directly related to the nature of his or her main business. If the client digs a commodity out of the ground, or comes by the material through the purchase of scrap or actual material, he or she then fixes a price by selling the commodity in the futures market.

What would a speculative account do?

MR: A speculative account would trade simply to cash in on price movement. The client takes a view of the market. If the price of gold appears cheap to a speculator, he or she would purchase a future position in gold or in options. If the market improves in price, the position would be more valable.

And they would sell if they thought that the market was going to go down. Let's talk about what type of service you extend to professional accounts, and what type of service you extend to speculative or retail accounts.

MR: We try not to discriminate. Any client of the firm who has fulfilled all our capitalization requirements will receive from us numerous services that will allow them to function. First and foremost, we will advise a client of market movement as it occurs.

Do you call the customer with any sort of price change, or do you ask the client to give you perimeters, to be called only when the perimeters are hit?

MR: We do both.

What other services do you provide?

MR: We tell customers what the overseas markets have done prior to our market opening, and also subsequent to our market closing. Sometimes the client will feel that he or she would like to start a position in Hong Kong because New York is already closed. Or, conversely, they may wish to start their position in London or Switzerland before New York has opened. We are now capable of trading most precious metals almost twenty-four hours a day, six days a week.

Is that service open to any client of the firm or only to certain types of clients?

MR: That service is open to anyone who is financially capable of being in those markets.

You and most full-service firms provide market quotation, price service, extended-hour trading capabilities. What other services do you provide a client?

MR: We have available to us up-to-the-minute market and raw material information from various sources. We are also aware of world-wide movements affecting the price of precious metals. Many people feel there is a direct relationship between the price of the dollar versus foreign currency and the price of gold. Similarly, there may be a relationship between the price of oil and the price of gold. Recently, as the price of oil has fallen, the price of gold has gone in the same direction. Historically, there's been a relationship between the price of gold and the price of silver. Sometimes these relationships expand and contract but they usually move in the same direction.

So you can provide research in addition to quotation service. How about other services such as financing for the metals themselves and for futures?

MR: We are happy to finance precious metals for our clients. Precious metals in the publicly-traded markets are the most readily marketable commodity in the world. Many people, known as gold bugs, are anxious to have their net worth involved with the price of metals. They feel that the long-term view for metals is easier to identify than the value of paper money. For that reason, there is a very good market for metals and, for the same reason, we are ready to finance precious metals for clients on a certain predetermined formula basis.

The client doesn't have to start by dealing on the recognized exchanges. He or she may buy the metal from a mine, bring it to us, allow us to have it tested and assayed, and then we would lend money providing we hold the metal as collateral. Or we hold a recognized warehouse receipt.

Are there any other areas in which you give service to a client?

MR: Recently, the whole world has discovered technical analysis. This is a map of historical price patterns that occur in futures. Many times there's a tendency for these patterns to repeat themselves. Therefore, we follow the research of our technical department and create some proprietary ideas with regard to tactical movements in the markets.

Do you find that professional clients require more advice and price information than speculative clients?

MR: It's a question of personality. Some professional clients don't want to have any comments and are only interested in prompt, effective market execution.

I've asked this question of floor traders, but I'd like to hear an upstairs broker's point of view. How long do you feel it takes for

an order to be executed and reported back to you under normal conditions?

MR: That's like asking me how long it takes to drive on the Long Island Expressway to my beach house. It depends on traffic. But we're happy with the knowledge that our orders are executed very promptly. Sometimes a delay occurs, not in the execution, which takes place within minutes or fractions thereof, but with the information coming back to us because of the crowded condition of the market and the phone lines.

But phone lines help most of the time. They give the retail client in Pittsburgh the same execution capability as the professional client in London. They also can use telex and, sometimes, mail.

Why would you want to transact an order by mail in this modern day of electronic communication?

MR: We may have a client who tells us to buy or sell according to a certain pattern; for example, to sell at every ten-dollar up movement in gold or to buy at every ten-dollar down movement. Some people buy so many ounces of gold every day right after the opening. These types of orders, once they're entered, don't need to be pursued except to confirm the information on a what-has-been-done basis.

Do you offer hedging programs, or professionally managed money in terms of precious metals? Do you extend a service that will naturally fit both a user of the precious metal and a miner of the precious metal so that you have a program for them to follow?

MR: That's exactly what I was alluding to when I told you that we confirm by mail. This is a usual acquisition and hedging program procedure. People want to do so much business every day. They look for average prices. Or they look for a long-term acquisition program if they're buying or a long-term sales program if they're mining and selling.

All right. Now let's turn the page around and ask you what you expect from a client?

MR: We expect a client to know his or her own mind, and we will help supply the information.

What do you expect from a client financially? Let us say, for example, the client initiates a position, and the firm requires a deposit on that position. How promptly must that deposit be paid to the firm?

MR: We know that a client who makes a transaction expects to receive the money when a profit is made within the settled period, and is very confused and disappointed if we don't live up to it. In the same way, we expect a client to promptly meet his or her obligations so that we can function in the way in which we have contracted to.

Do you expect a client to know the nomenclature of the business and place orders in a particular way with you?

MR: We have a very simple system. We believe that a client doesn't need any formal training or any licenses in order to do business with us. We will explain to the client what our role will be and, on each transaction before entering the order, we will always repeat it until we are positive he or she understands what we are being asked to do. At that point, we'll carry out the transaction.

Do you have any advice for someone who want to start trading in precious metals?

MR: Yes. Know what you're looking for in the market. Decide when the market has moved enough to let you take your profit, or, in adverse circumstance, at what point you'd like to step out of the picture and take your loss. One of our biggest problems among professional as well as speculative accounts is the question of ego. Once the market goes against a person, the usual tendency is not to look at it objectively. The only way to be objective is not to be in the market. Once you become involved, it's very hard to really see what is haappening.

I'm sure you've heard the saying that speculators always lose and professionals always make money. Do you feel that is so?

MR: I believe that, in trading of futuures, every day someone will lose and someone will win. The fact that many speculators have lost may be a confusion in terminology. We have to deal with the number of dollars involved and not with the number of players in the game. If you have fifty speculators all of whom have lost ten thousand dollars, that's equal to half a million dollars. Many professionals lose half a million dollars in two days. Therefore, you'll have a statistic that fifty lost or one gained but the numbers don't really tell you the story. You have to look at the dollars involved, because that's what we're here for.

Are there any other services you provide that could give a client a special edge?

MR: Sometimes we find little aberrations in the market, which make it interesting to participate indirectly. Specifically, instead of buying gold, we might recommend that the client who buys gold sell platinum. The price of one material might outrank the price of another material. Some time ago, gold was selling at $150, maybe 30 percent higher than the price of platinum. This as an historical aberration, and anybody who bought platinum and sold gold made money even though the overall precious metal market declined.

Do you monitor your accounts on a daily basis?

MR: No. However, our computers flag any fluctuation resulting in a status change and we promptly inform the client.

Any other advice?

MR: Nobody should enter a market for the first time.

What do you mean by that interesting statement?

MR: You should investigate before you invest in futures in the same way that you investigate any other purchase that you make.

SEVEN

Clearing an Order—from Broker to Clearing House

A favorite pastime of many floor brokers is debating the ideal set-up for expediting orders and receiving execution reports. With apologies to the vendors of state-of-the-art electronic systems, I have found over the years that the telephone is the most reliable and quickest means of communication. All orders must be written, numbered and time-stamped as soon as they are received. Buy orders should be given even numbers and sell orders odd numbers to avoid a mix-up between buys and sells as they are reported. This also will flag obvious errors. At the end of each day, every order must be checked and repeated by telephone between the upstairs order desk and the floor station.

A well-organized order room is just as important as a well-organized research department. A floor broker should ideally have at least six telephone order stations per firm that is to be serviced provided the volume warrants it, so a full service telephone requires six turrets and twelve clerks with two people continually manning each station. A taping device should be attached to each phone, not only to pinpoint errors and fix responsibility, but to analyze order flow.

Teleprinters should be used for orders that are not immediately executable and for recording executed trades. Teleprinters communicate reports in writing between the floor and the various order rooms and a copy is automatically sent to the margin department. What this means to a busy producer can best be illustrated by a firm that is able to provide a complete run of each account before the end of the day with a customer's updated position by transaction. Today, a modern account executive in a major firm can recall a customer's account with the latest position tabulated.

The distance between a phone booth and floor broker is a crucial concern. Many commission houses maintain booths and telephone order stations that are physically too far away from the broker on the floor who handles their business. Consequently, a messenger may have to fight his or her way through a busy crowd with an order, inadvertently causing financial losses because of the delay. Further, even if a broker executes an order in a timely fashion, a messenger may be delayed in bringing the report of the "fill" back to the telephone order station. If the market has changed before the report is in, the customer might get the impression of a bad execution.

Seller Meets Buyer

When a seller's broker and a buyer's broker complete a transaction, the exchange has to "clear" (that is, the paperwork has to be completed) for both sides of the trade. In other words, every buy is matched with a sell order and the trade has to be recorded in the proper clearing member's account. A mismatched transaction can turn into a costly error.

Clearing members have their accounts updated daily, and have to provide funds or receive funds at the exchange for liquidated transactions.

The exchange may also ask clearing members for "intra-day variation deposits," known as margins. Minimum margins, or deposits, for every futures contract bought or sold are set by the exchange's board of governors. The money is a "good-faith" deposit to ensure that a member would make good on a trade. Customers' profits or losses, based on the daily settlement price of each contract, are held by each clearing member, not by the exchange.

However, the clearing house of an exchange stands behind every cleared transaction. In order to become a clearing member, a candidate has to provide evidence of financial integrity and meet capital requirements.

Pat Thompson is a NYMEX floor trader, and, until recently, was Compliance Director of the NYMEX. He is an attorney and former government investigator and prosecutor.

Would you agree that a good compliance department promotes confidence? Can you attribute the increase in level in recent years to stronger confidence by the public in integrity of the market and the system by which orders are handled at the exchange?

PT: I think the relationship between compliance and volume is one of the great intangibles, but I think it's proven to some degree by the customer participation level today versus the participation that existed, say, three or four years ago. That was shortly after the CFTC had been formed. [The CFTC is the Commodity Futures Trading Commission, the governmental agency that regulates the futures industry.]

In 1984, the CFTC's tenth year, it is highly successful and vigorous in its policing actions. The commission is unique as a regulatory agency, since it encourages a "free" and "fair market" system in which price is determined by true underlying economic variables in a particular marketplace. It uses its full regulatory powers only in those instances where forces within the system subvert the "free" and "fair market" principle.

PT: I think it's clear that customers will not trade contracts or instruments if they believe they are not being treated fairly in the execution of orders. To draw customers to a market who will trade that market regularly, the exchanges must give them the feeling that they are getting a fair shake. The compliance department of an exchange is their insurance. A compliance department complemented by a strong and independent enforcement program probably attracts trades to the market.

Do you use computers or any new methods for tracing trades?

PT: In 1982 we acquired a computer capability. But let's start at the beginning. The way in which we match trades and clear them gives us a leg up on most every exchange.

When you say "matched and cleared," you mean that if floor broker Smith is selling something and floor broker Jones is buying something, those trades have to be compared and must coincide by the end of the day to be labeled bona fide trades?

PT: That's right. Each trade that occurs is compared by computer to make sure that Broker A, if he or she is trading with Broker B, will be matched on a sheet and cleared for exchange purposes as the executing brokers on a particular trade. Our computer system automatically will match the trades on the Street Book.

What is a Street Book?

PT: A Street Book is unique to the NYMEX. It is the computer-generated document that will show the following matches: the clearing member that Broker A is executing for and the clearing member that Broker B is executing for; whether it was a customer trade, a house trade, or a personal trade for both sides of the trade; and the order number for both sides of the trade. So, when the Street Book is given all that information, we already have a fairly good idea of what the flow of the market is, the order in which trading occurred, and the identity of the persons trading. The Street Book also allows us fairly quickly to obtain any type of back-up documentation (such as order tickets or account statements) that we might want in order to audit a trade. All the information is indexed and available to us in the Street Book.

Now, we added to our capability in 1982 by acquiring a computer system that allows us to mesh that information and call off just those parts of the Street Book that we might be interested in. We can put into the computer directions such as: "Please give us all trading for Broker A in platinum and palladium over any period of time." You can try to get as much or as little information as you want, and to hone that information down and make it as close to the type of analysis that you want, and have the computer do the entire analysis for you. Once you take a look at that isolated information, it gives you an idea of whether or not you should go further in your investigation. So I think it's been a really good aid, one that helps to prove cases to some extent, but mainly that helps save us so much time in the preliminary investigative work.

When you say prove cases, you mean when you begin an investigation of trading?

PT: It allows us to save a lot of time at the initial stages because it'll tell us at least whether or not we have a reason to proceed. And then, it helps us to do other investigative things, such as obtaining the underlying documentation.

Pat, let's get specific about compliance issues. Let us assume that an order good for the whole day was placed with a brokerage firm to buy a contract of platinum at a price of $398 per ounce. The range for platinum that day was $400 high and $398 low, but at the end of the day the account executive informs the customer, "You were unable to have your order executed. You didn't buy it." Does the customer have a complaint, and, if so, what can be done?

PT: Okay, at that point we don't know whether or not the customer has a complaint, but we can investigate whether there was a poor reason for the order not getting filled. Then we ask the customer or the clearing member (the brokerage firm the customer's order went through), or both, to just give us a ring to let us know that they have a complaint so that we can do our best that day to at least keep the documentation in house. That is because with our current volume the documentation sometimes gets sent to the warehouse practically the same day the trading occurs. Then we ask for a written letter explaining the circumstances and asking us to look into the matter and investigate the circumstances of the trade. Once we receive that letter, we will open a customer complaint file, and will respond to the customer acknowledging the letter, explaining that once we have completed our investigation, we will release as much public information as we can concerning the investigation. If a disciplinary action is commenced, the customer would be informed and we would enclose the arbitration rules of the exchange. In order to press a claim in terms of receiving money back—if the customer feels that the money can't be recovered—then arbitration judgment is possible.

Once a customer file is opened, we treat it as any other member complaint or any other self-generated inquiry from inside the compliance department. We will collect all the appropriate documentation, such as order tickets, pit cards, the Street Book, and purchase and sale statements. In reviewing the Street Book, we might find other suspicious trading activity. We will look into that because it could explain why this trade went unexecuted, although that's certainly not the issue at point. And so we will initially just try to get a lot of information and conduct interviews with the brokers to determine what the market was like at the time this order was executable, who was handling the order, whether or not the order was represented in the ring, who the other active traders were, and so forth. We, of course, try to determine whether or not there were any conflicts of interest, such as personal trades by the brokers on the account. Once we have acquired as much information as we can, the investigator and the manager of trade surveillance analyze the data. They then make a recommendation to the head of the compliance department as to whether or not a violation has occurred. They may feel that either the evidence is insufficient or that there is, in fact, no basis for the complaint.

Does this mean that any customer of any firm anywhere in the world that is a member of the exchange (or even a non-exchange member whose broker clears through an exchange member) has a right to call upon the exchange for help?

PT: Absolutely. Our investigators are trained to treat a complaint the same way as one from a member.

In a very hectic market, does the volume of complaints greatly increase?

PT: I would say so, yes. There are more complaints because there is more uncertainty in the contract pricing that's been done during the day in the trading. Also, a greater volume of customers trading means that there will obviously be more people who are unhappy with their trading results. The other factor that seems to increase the number of complaints is price volatility, where markets often move in many directions during the day. The more volatility there is, the more customers will be asking for [an order] fill because they feel that the market may have hit their limit price or their stop price several times. If they feel they have more opportunity for a fill, they will want to have a missed fill investigated.

Do you investigate trades even when you don't have a complaint, just as a normal spotcheck procedure?

PT: Yes. As part of the daily routine, each investigator has to review a certain segment of the Street Book every day. Each concentrates on a certain trading segment of the day or a certain commodity for a certain segment of the day to review for any questionable activity. Each investigator will also have daily floor surveillance which he or she engages in, and at times might hear rumors that could lead us to use our computer to generate trading activity for, say, the previous week of a particular broker. So there are many ways in which it's done, but generally there is the daily routine as well as the computer capability that we use to monitor trade.

Do you keep trading records? If so, for whom and for how long?

PT: Yes, under CFTC regulations we must keep records on all registrants—meaning brokers on the floor and clearing members (the brokers who deal with the public generally). The exchange must maintain all records of trading for a period of five years after the date of the transaction.

Do you have any words of advice for eager clients entering the metals markets about keeping records and proper trading conduct?

PT: Well, the first thing is to utilize the firms that are recognized leaders in metals trading. Try to contact customers and traders who actually do business for and against these firms. At least there is some security in knowing that people who do business on an ongoing basis in the metal industry feel that a broker has a good, sound reputation.

As to record-keeping, I would say, just be a pack rat and don't throw anything away. Make sure that you keep good files and that you keep them up to date. Also, any time you have a complaint or a question about a transaction, it's easier to lift up the phone and call an exchange, so that investigators can take a look quickly, than to wait until memories fade. In commodities, markets change so quickly that the sooner you ask your questions and get answers, the better the results. Sometimes we can handle an inquiry in five minutes; sometimes we have to do a little digging, but the closer we are to the event in question, the better the results.

Do investigation departments deal with both exchange-related complaints and non-exchange, commodity-type complaints?

PT: Exchange-related complaints are defined as any complaint related to a transaction that occurred on a recognized contract market. We no longer deal with other complaints that relate to commodity trading done either through leverage dealers or through commodity options dealers and the like.

Is there a problem in the public dealing with entities that are not members of a recognized exchange? These firms don't really fall under anyone's compliance procedures, though some state regulators are getting tougher. But are your hands tied when it relates to non-exchange complaints?

PT: That's right. And when I was with the CFTC (the commodity industry regulator) I'd say those types of complaints outweighed exchange-related complaints ten to one. I think it's just a fact of life that if a firm or individual has gone to the expense of buying a seat on an exchange and paying the business costs involved in setting up a firm to do trading, they will be more careful because they have a tremendous financial stake in staying in business by complying with the rules. Off-exchange firms, as a rule, are prone to engage in unfair tactics and may frequently violate rules we think are important.

EIGHT

The Computer Craze and Technical Trading

In 1974, four major New York commodity exchanges consolidated, sharing a common trading floor at number 4 World Trade Center. It became known as CEC, the Commodity Exchange Center. A number of computer companies also moved into the building and began offering services to floor traders. As the popularity of computer-inspired trading programs grew, many floor traders found that they could no longer ignore the influence of these computer-generated orders.

Futures commissions merchants (FCM), brokerage houses that trade commodities for clients, started to attract large amounts of public capital which were placed into commodity pools. In a commodity pool, modest amounts of capital from individual investors are added together to form a large pool, which is then managed by a trading adviser using computer programs. Commodity pools are often called the mutual funds of the commodity industry. Many major industry figures feel that investors with modest capital belong in these professionally managed account programs. Due to the explosive growth of the industry there are many more funds today than before.

These pools, or commodity funds, buy and sell precious metals based almost solely on computer program instructions. One cannot go very far on the floor of any commodity exchange today without tripping over a pile of computer printouts.

Gold and Technical Short-Term Trading

The difference between making price decisions based on fundamentals as opposed to technical considerations is significant. Professional gold research analysts differ in their approach to the market. Some say that economic fundamentals are the only major and worthwhile source of information; others feel that the only answer is the charting of "technical considerations." Charting is very popular in the American futures markets, and brokers adapted it to commodities markets when they came into this field.

Charting is a relatively easy way of attempting to forecast short-term price moves. However, charts can be very misleading. For example, let's look at the effects of the Federal Reserve on curbing inflation. Say that signs from the Fed begin to show that fundamentals are slowly changing, and data from the Fed and the U.S. Commerce Department show that inflation is starting to come under control. If, in fact, inflation is coming under control, and at the same time interest rates are slightly higher than before, gold certainly becomes less attractive to hold (except for investors from nations with a political problem or a major currency devaluation). While you may have a bullish chart, it may not help, because the bullish chart may be met by resistance from professionals. In fact, while the chart is bullish, the fundamentals could be bearish.

On the other hand, one cannot ignore the so-called "technical indicators" that charts provide. Making and interpreting charts is based on a complex science of price behavior, and most of those using charts in today's gold markets follow certain rules that are incorporated into computerized programs. There are many good computer services that will provide all kinds of chart patterns, chart explanations and even chart forecasts of price behaviors.

The most important thing to remember about charting is that it influences markets. Most participants in a market will maintain price levels based on what they think people who follow charts are being told with regard to the time to enter or leave a market. Thus, relying on charts has tremendous drawbacks. The futures market has millions of chart followers, all of whom purchase when there is a chart breakout in the resistance line, which generally constitutes a new high. Precisely because so many have acted at the same time, it can only have one effect—the price will go up. With charting, everybody is trying to go through the same door at the same time in a crowded room.

The best way to define charting is to say it is the history of the price behavior and volume behavior of the underlying commodity. Chartists break down their approach to attempt to show the price behavior of the high and low of the day, high and low of the month, high and low of the year and high and low of the life of the contract. Some say that two and three point reversals are necessary; others say that lines must be drawn in order to show certain formation between highs and lows called bottoms and tops, heads and shoulders, flags and many other catchy names to describe certain chart patterns.

Charts or fundamentals are not the only key to any market in my experience, but in over ten years of trading precious metals as a floor trader I have come to believe that a combination of everything, including one's nose and one's feelings, are an important part in forecasting price behaviors.

It must be remembered, of course, that a key component of futures trading is market psychology, and temporary market psychology can always distort longer-term trends.

Arthur Dock is a long-time member of the financial community. He is the ring chairman of the floor committee of the NYMEX metals area and his charting service has been used by many floor traders over the years. He is a convinced technician and active floor trader for his own account.

What can your charting method do?

AD: It can teach a person a technique by which he or she can go into any commodity any time of the day. They can use the CRT (video screen) to find out what trading reversal my chart is showing—whether there's an upside reversal or downside reversal—and then can immediately start trading.

During the course of a day, do you keep a small chart or a copy of your chart in your pocket and add all the price movements, or do you just update at the end of the day?

AD: I update at the end of the day, except for platinum and gold, which I trade in the pit. I keep a running chart. At all times; I know where the major and the intermediate resistances to support areas are. Therefore, I know where to take my long-term position, my overnight position. During the day I trade only my trading reversals.

You mean, if during the day the market continues in one direction, you will continue your program in that particular direction.

AD: Exactly. I will ride with the market.

So you are a proponent of "let your profits run and cut your losses right away"?

AD: That's exactly right. I am like Sir Isaac Newton. I know that things in motion tend to remain in motion. And I follow that motion.

Do you feel that precious metals generally follow a particular direction for a particular amount of time?

AD: No. I am not interested in time. An upside market in gold or platinum will continue to follow the uptrend line. When it breaks that uptrend line, my long-term position will be bearish.

At one time you were requested to publish your charts. How long did you publish your service?

AD: I published it for five years. But I stopped because I became lazy, and I decided to have an easier life. I used to work three, three-and-a-half hours every night to publish my paper every morning. I got calls from my clients at night asking me to advise them on the action so they could trade in Hong Kong, in Australia, in London. I had no life for myself.

Coming back to your daily activities in the ring as a trader for your own account, when you come in in the morning, do you generally participate at the opening price? Or do you wait to see how the market trades before you become a more serious participant?

AD: I will usually wait, unless prices hit my resistance or support area.

You watch major participants in the ring to gauge prices, don't you? Do you look at a commission house, a trade house, or is it a group of participants that you watch?

AD: Usually it will be a trade house because the trade, even in an up market, will constantly feed into the up market, whereas the commission houses issue their orders in bulk and then are done.

Do you generally feel that most commission houses act on one side of the market, and most trade houses act on another?

AD: All trade houses will normally act on one side of the market. A commission house will act on both sides of the market depending on what their account executives tell them to do. A commission house may have trade clients as well as other types of speculative clients. They don't really show market direction.

Is there anything that you feel might be important for someone to know who has not been a participant in the metal markets?

AD: Yes. If you are going into the commodities business, make it a business. It should not be an outlet to gamble because you can get creamed. If you go into this business it can be extremely lucrative, but you must first learn the business.

How can newcomers do that?

AD: Take the course give by the New York Mercantile Exchange or the New School for Social Research or the New York Institute of Finance. Read any books from which you can learn the basics, and then learn the basic charting techniques. Then observe on the floor—and remember that you cannot observe long enough. If I had been in this business eighty years, I would still be learning because new things happen every day.

What is the most interesting experience you've had in trading for your own account on a daily basis—in terms of chart price movement? Was it influenced by a news event, an economic event, a political event?

AD: The most dramatic thing I've ever seen was in the spread market, believe it or not. (See glossary for a definition of spreads.) People trade spreads for an assumption of safety. They don't want the exposure of having an outright position. But I have seen platinum spreads go from \$40 to \$ −30 or \$ −35. The October contract was selling for \$40 above July's. During the course of trading, about a week or two, the price for October went to about \$30 or \$35 below the price for July.

What cuased that chart differential?

AD: There was a shortage in spot metal. Physical metal. The only way you could get physical metal was by paying a premium. Therefore, you were sacrificing the price of the future. So, they paid the premium for July and the future went to $30 or $35 *below* the spot.

So, what you experienced was the inability of sellers to make a delivery in the short period of time allotted to them. Therefore, they had to repurchase the contracts that they had sold on the exchange, changing the premium differential between the July and other future months.

AD: Not only that, but the user, the producer, needed the physical inventory to continue his business and had to buy spot platinum at premium *then,* at any price.

So that economics also play a very large part?

AD: No question about it.

Dennis Turner is a well-known computer specialist in commodities trading. He has been programming and testing trading systems for 15 years. He has worked for I.B.M. and has written two books, *Trading Silver Profitably* and *When Your Bank Fails.*

How did you develop your trading system?

DT: I have tested about 5,000 different trading methods in precious metals on a computer. A computer program is a set of instructions that tells the computer to manipulate data. The manipulation can be inputting data in files, doing arithmetical operations, manipulating prices, etc. In other words, a computer program will take in a set of gold prices and it will add and subtract gold prices from one another and come up with a third number. Furthermore, the program will create a set of rules which test what would happen if you bought or sold gold under certain conditions.

The easiest example of a program is a moving average system. The computer program will add 10 numbers—10 days' prices—and it will create a 20-day moving average. Each day forward it will add and subtract to create the 10-

day moving average. If the moving average rises, it will signal you to buy; if the moving average falls, it will give a sell signal. You can do this also with "channel" methods; that is, if there's a thrust up above a recent high or down below a recent low. Or it could be exponential averages, or even more complicated methods.

What are the elements of the daily movements that you need to program into your computer system?

DT: Most people who are programming technical methods use the open, high, low, and close of each day. Many also add volume and open interest figures. So there really will be six pieces of data for each day. If you're testing five years' worth of data, say 250 days a year, you'll come up with 1250 days times six, which is 7,500 pieces of information.

How do you get that information to put into your computer?

DT: There are many ways. The most difficult is by typing it in out of old *Wall Street Journals*, but that way is obsolete. You can buy the data on floppy disks from many companies. There are very complete data bases on commodities and you can write programs on machines, or you can extract the data onto you floppy disk for your home computer—for a price, of course.

If people wanted to initiate a computer program or start to follow precious metals through computer programs, what information would they need and how would you advise them to start out?

DT: I would advise them to start out by buying a few books on commodities, particularly those that discuss technical trading methods, in order to learn technical trading methods. They need to know how to make the standard calculations of moving averages, price channels, momentum, oscillation and a few other things. Then they would need a computer that has a 128K memory, a modem, a describer tube—preferably two—and a printer. It's really not a very complicated system. The most important thing is to get a good communications program that allows you to communicate easily with services that provide data to personal computer uses.

Then you would have to open up an account with one of these companies, which is quite easy—you fill out a form and send in $50.00 or something like that. Now, the company will usually provide the information, in their manual on how to extract the data or how to use the data from their system. The manuals for the communications program are very explicit, and usually these communications programs are very easy to use, and you get abstract data.

Of course, you have to know how to program it. Programmers who do technical trading methods usually program in BASIC. BASIC is very easy to learn, but I would suggest that if you are in a hurry you do not try to learn it from a manual. Most companies that sell personal computers now also provide courses in BASIC. They may charge you $200 or $300, but in a few weeks you can program in BASIC well enough to do your work.

Can you estimate the percentage of daily trading that is directly traceable to computer methods?

DT: That's not easy. It's certainly less than 50 percent, probably somewhere between 10 percent and 25 percent.

Do you feel that computers will have an increasing importance in trading precious metals?

DT: Yes, but when great international events occur, or super inflation occurs, most buyers of gold are not going to wait for computer signals. In those times, I would guess that most of the buying would be fairly panic originated. But when these factors are absent, the amount of trading that is at least partly dependent on indicators produced by computers would continue to rise.

Does computer tracking of other metal futures relationships, which many call "ratios," provide any clues?

DT: Sometimes yes, but you can't fall into the fallacy of believing that merely because something is called a precious metal it will necessarily do the same thing as gold.

When do you input your data?

DT: The computer data base I use is updated every day. My programs use the open, high, low, and close as well as the volume and open interest figures.

Do you track the relationship between metals?

DT: I don't use the ratio between silver and gold very much. I look at it mostly for interest. I assume that it is statistically unreliable. The ratio of gold to silver can change and stay changed for very long periods of time and, consequently, if one spread between the two based on a frequency of idea of what the ratio should be, one could lose money for years at a time. It's not statistically reliable, and therefore I don't use it.

After you update your computer, does your finished product give you parameters for buying and selling?

DT: Each morning I get buy and sell signals for all major markets.

You probably hear a lot of floorbrokers' talk that there are computer stops at various levels in the market.

DT: You hear a lot. Stops are so-called chart points that most traders become aware of by subscribing to any one of a number of computer charting services that show when large pools of money or computer-directed funds tend to buy or sell. Normally, this takes place on new highs and new lows in any 21-day segment of the move.

Should everyone use the same programs for trading?

DT: No. Because if everyone, or a lot of people, were trading at the same price, we would get large error of execution.

What do you mean by error of execution?

DT: Simply, if ten people were using the program, it would cause the value of the program to deteriorate because the prices at which the orders are filled would be much worse than if only a few people were using a particular system. And this is one of the downfalls and problems of a large firm using a system. If a Merrill Lynch-type firm has a system and they had an order for 1,000 lots, they will distort the market by executing it in one fell swoop.

Trading Method

The trading method we use contains components of both a trend-following method and a swing, or thrust, method.

Many trend-following methods correctly indicate the long-range direction of a market. Examples of such trend-following methods include *moving averages* and *channels*. A moving average method averages closing prices over a period of time—for example, the last 10, 20, 30 or 50 closing prices. This is intended to weed out temporary fluctuations and to calculate a reasonably representative price over the specified number of trading days. If the moving average rises, the method predicts the market is in an uptrend, and a long (buy) position is taken. If the moving average declines, a short (sell) position is taken.

There are many variations of moving average systems. One might specify a short position if today's moving average is below yesterday's; another will specify a short position if the closing price today is below the moving average. A third may employ two moving averages; if the shorter is below the longer, a short position is indicated.

Channel methods define a high boundary and a low boundary over a period of time, which might be 10, 20, 30 or 50 trading days. The channel boundaries might consist of the highest and lowest traded price within those trading days. Price fluctuations within the specified trading days are said to be random fluctuations, signifying nothing important. Only when the market moves outside the trading channel does the trader believe that a significant move occurred which indicates an important change in price expectations. If the price moves above the upper channel boundary, a long position is taken. If the price moves below the lower channel boundary, a short position is taken. Effective parameters can only be discovered by testing. A ten-day moving average may produce losses, while a twenty-day moving average may produce fantastic profits. There is no convincing *a priori* reason why any particular number of trading days should be effective. Only by testing can effective, profitable parameters be discovered.

We have tested hundreds of parameters of many different versions of moving average and channel methods, as well as many other methods not discussed above. In the end, we selected the trend-following method which we believe proved most profitable.

Trend-following methods alone are not adequate. Because they make projections based on a large number of past trading days, they are often late, entering the market after a substantial move has occurred, and not exiting or reversing the position until much of the move has been reversed.

For example, gold may move from $400 per ounce. A 20- or 30-day channel or moving average method may not enter the bull move until gold reaches $430 per ounce. It will stay long throughout the entire move up to $500 per ounce, but will not liquidate or reverse the long position until the market has fallen back to $460 per ounce. A $30 profit would have been taken on a $100 move. The results could be worse, and often are. The long position may not be taken until $450 an ounce, and kept until the price has decline from the high all the way down to $460 an ounce. Only a $10 profit would have been taken.

Another weakness of trend-following methods is the large losses taken when wrong moves are predicted—even in a version that produces large profits in the end. A trend-following method may enter gold at $450 per ounce, and then the price may erode. Because the moving average declines slowly from day to day, as today's closing price must be averaged with the 9 or 19 or 29 previous closing prices, a reversal or liquidation may not be signalled until the price declines to $410 per ounce. This is a $4,000 loss. Several consecutive losses may occur. This can be psychologically devastating. The method may successfully catch a $10,000 move and an $8,000 move, thus producing a profit even after four or five substantial losses. But traders are likely to discontinue using the method if the losses come at the beginning.

Yet trend-following methods, with the proper selection of parameters, are usually successful in indicating long-term price direction. A good trading method will take advantage of the direction of price moves indicated by a trend-following method.

In our trading method the long-term trend influences the calculation of the stop and reverse points, along with the thrust or swing method, so the influence of both determines when our trading method buys and when it sells.

The price behavior inducing us to enter or exit a position is otherwise not determined by the trend-following method. Instead, we employ a short-term thrust or swing system.

Our thrust system looks at two features of short-term price behavior—the amount of fluctuation in price movement, and a very recent representative price. We will not reveal the exact construction of either. Our measure of fluctuation indicates how much price movement can be expected in normal up-and-down daily trading, and how much is signficantly more than expected. We measure the price movement each day against the measure of "normal" price fluctuation, based upon the recent movement of prices. If the price moves sufficiently from the short-term representative, or base, price, we predict that a swing has begun, which will continue in the same direction.

If the current position is neutral, we take the following action:

1. If the thrust is the direction of the long-term trend, a position is entered. For example, if the thrust is up and the long-term trend is up, a long position is taken. If the thrust is down and the long-term trend is down, a short position is taken.
2. If the thrust is contrary to the direction of the long-term trend, we remain neutral. For example, if the thrust is up and the long-term trend is down, or if the thrust is down and the long-term trend is up, we do nothing. No position is entered.

If the current position is long, we take the following action:

1. If the thrust is down and the long-term trend remains up, we liquidate the long position. If the thrust is down and the long-term trend also turns to the downside, the long position is reversed to short. Of interest is the fact that the long-term trend rarely changes on a thrust, so most positions are liquidated, not reversed.

2. If the thrust is up, we ignore it, since we are already long.

If the current position is short, we take the following action:

1. If the thrust is up and the long-term trend remains down, we liquidate the short position.
2. If the thurst is up and the long-term trend also turns up, the short position is reversed to long.

The thrust or swing system works on its own. Why then is it used in conjunction with a long-term trend system? A thrust is more likely to indicate a continuing swing in the direction of the thrust if the long-term trend agrees with the thrust. This isn't a hypothesis, or something which sounds as if it should work. It is the result of testing. The percentage of profitable trades increases when the thrust method is used in conjunction with a trend-following system.

When a thrust contradicts a long-term trend, it is less likely to be correct. Therefore we liquidate instead of reverse. Reversals are more likely to be wrong if they move against the direction of a trend. So it is more profitable to liquidate. This reduces the probability of a whipsaw, or losing trade, which would likely occur if the position were reversed on a thrust against the trend.

Technical Traders' Glossary

BAR CHART:	A daily chart of the high, low, and close prices; and volume and open interest for any contracts.
CONTRACT AVERAGE PRICES:	A display of the five-day moving averages for all active contracts, as well as the five-day moving average for all expired contracts in the same period.
MARKET QUOTES:	A display of the daily market quotations for all contracts of any commodity.
MOVING AVERAGES:	A display of the 5-, 10-, 20- or 40-day moving average for any contract.
POINT AND FIGURE CHART:	A daily chart of the last (up to sixty) reversal points for any contract.
PRICE VOLATILITY HISTORY:	A display of the price volatility of any contract in previous years for the same time period.
RECENT PRICE:	A display of all or part of the price data for a given contract.
SPREAD:	A display of the spread between any two contracts which traded on the same day(s).
SPREAD CHART:	A daily chart of the difference between the closing prices of two contract.
SPREAD HISTORY:	A display of the history of a particular spread from the current to the most distant contracts.
WEEKLY RANGE:	A display of the highest high, lowest low, and last price on a weekly basis for a given contract.

Tom Petito is director of the Trade and Reconciliations Department at Comex. He has twenty-five years' experience in the commodities industry. A former broker and limited partner of a futures commission merchant, Tom has compiled a glossary of essential commodity terms (see pp. 129-37).

How did you compile your glossary?

TP: Some of it was compiled from exchange booklets, some from books that I've read, and some from just listening to somebody saying something lucid. I verified things, and just took the terms that I felt helped me to learn the business. I put them down on paper so somebody else could learn it as well as I learned it.

Well, Tom, in effect you've performed a public service. Do you think this will someday be standardized?

TP: I think something like this is overdue. We teach clerks who work for brokers. They've told us that they've learned more in their one week with us than they've learned in a year or two of being on the floor, because the floor population is too busy to teach. So something like this glossary would help everybody out.

Is there anything that you feel would be helpful to the general public to supplement your glossary? I mean aids in addition to "how to" books, and such required reading as The Wall Street Journal, Journal of Commerce, Business Week, *and* Fortune?

TP: There is a course that could be helpful—the Futures Industry Association training course. It offers courses in Chicago and New York, and has a correspondence course that gives a lot of detailed information about the entire business. (The Futures Industry Association, headquartered in Washington, D.C., can supply a schedule of its courses.)

How should a new broker enter the exchange after passing the membership committee and acquiring a seat?

TP: I feel that any novice should go through some sort of a training period to let him or her know exactly what the regulations are, how to trade what happens when a trade busts (does not clear properly) or doesn't match, how to clean it up or rectify an error, etc. Many of these new brokers rely on their clerks. There's also a manual that's put out by the Chicago Board of Trade that goes into varied detail on the entire business.

What is your main job at Comex?

TP: Well basically, I watch for any transaction problems that might occur on a daily basis. A transaction is every individual line of brokerage, whether it be one contract or fifty contracts. So each line of brokerage to me is a potential problem. It has to be cleared, so that every buyer and seller is matched. I love to see loads of contracts done smoothly. But, as director of the trade and its reconciliations department, my area basically handles all the problems, or unmatched trades, that occur. One of the busiest times I can remember was February or March of 1983, when we did some tremendous volume. We were doing 60,000 to 65,000 transactions a day, and at that time traded only three contracts—gold, silver and copper. We were trading gross dollar value in the billions of dollars.*

*Several exchanges today trade billions of dollars worth of financial futures contracts on a routine basis.

Glossary

The following is a list of terms, as generally used in the commodity business. An asterisk (*) highlights the most frequently used terms.

Acreage Allotment—The government's limitations on how much acreage each farmer can plant on some basic crops.

Actuals—The actual physical commodity underlying a futures contract.

Afloat—Any commodities that are imported and exported, while they are on board ship en route to their destination.

Arbitrage—The simultaneous purchase of a future in one market against the sale of a future in a different market, in order to realize a profit from a difference in prices.

Backwardation—A market condition in which prices are lower in the future delivery months than in the nearest delivery month.

Basis—The difference between the cash price and the price of the futures.

**Bear*—One who believes prices will go lower.

**Bear Market*—Any market in which prices are in a declining trend.

**Bid*—An offer to buy a specific commodity at a specified price.

**Broker*—An individual who is paid a fee or commission for executing the buy or sell orders of a customer.
- (A) Floor Broker—a person who executes someone else's orders on the floor of an exchange.
- (B) Account Executive or Sales Representative—a person who handles customers and their orders in a commission house office.

Brokerage—The fee charged by a broker for the execution of a transaction.

**Bull*—One who believes prices will go higher.

Bullion—Gold or silver in bars or ingot form, which must be at least .995 fine.

**Bull Market*—Any market in which prices are in an upward tend.

Buy-In—An offsetting purchase to cover a previous sale.

Buying Hedge—Buying future contracts in order to protect oneself against possible increased cost of commodities that will be needed in the future.

**Call Option*—The type of option which gives the purchaser the right to buy the underlying commodity futures contract at a particular price any time between the purchase date and the expiration of the option.

Carrying Charges—The costs involved in owning commodities over a period of time, such as storage, insurance and any interest charges.

**Cash Commodity*—The actual physical commodity, sometimes called "spot commodity." See also *Actuals.*

Chartist—One who uses either graphs or charts in order to try to analyze the trends and price movements in the markets. A chartist is known as a "technical" analyst.

C.I.F.—Cost, insurance and freight, which is paid at the destination and included in the price quoted.

**Clearing House*—A central agency set up by the exchange, or authorized by it, through which transactions of members of the exchange are cleared and all financial settlements effected. It also collects margin from members.

**Clearing Member*—A member of a clearing association. Each clearing member must also be a member of the related exchange, but it is not necessary for a member of the exchange to be a member of the clearing association.

Closing Range—The range between the highest and lowest priced trades during the closing minutes of a trading session.

Commission—The money paid to a broker for the buying or selling of commodities in the futures or cash markets, and for otherwise servicing account.

**Commission House*—Any firm that buys or sells the actual commodities or futures contracts for customer accounts.

Commission Merchant—Any firm that buys or sells the actual commodities or futures contracts either for another member of the exchange or for a non-member.

Contango—A market condition in which prices are higher in the future, or distant, delivery months than in the nearest delivery month.

**Contract*—A term used to describe a unit of trading for a commodity future.

Contract Grades—The various grades of a commodity that are stipulated in the rules of an exchange as being deliverable against a futures contract.

Contract Month—The month in which delivery is to be made in accordance with a futures contract.

Contract Unit—The actual amount of a commodity that is stipulated in a given futures contract.

Cover—The purchase or sale of a contract to offset a previously established position.

Current Delivery Month—The futures contract that matures and becomes deliverable within the present month; also called "spot month."

**Day Orders*—These are orders that expire at the close of a day's trading. If they are not filled during the day, they are withdrawn.

Day Traders—Commodity traders who take a position in commodities and liquidate them prior to the close of the same trading day.

Delivery—The tender and receipt of the actual commodity, in the form of a warehouse receipt in fulfillment of a short position in futures during the period specified by the futures contract.

Delivery Month—A specified month in which the actual delivery of a commodity can be made under the terms of a futures contract.

**Delivery Notice*—This is the notice given by the seller of his or her intention to make delivery against an open short futures position on a specified date. (See also *Notice of Delivery.*)

Delivery Point—The area specified in the exchange rule book in which the delivery of a physical commodity can be made.

Delivery Price—The price established by the clearing house at which deliveries on futures are invoiced, and also the price at which the futures contract is settled when deliveries are made.

Discretionary Account—An account in which the holder gives power of attorney to another to make buying and selling decisions without notifying the holder.

Equity—The dollar value of a futures trading account, assuming it is liquidated at the going price of the markets involved.

**Exercise*—The conversion of an option by the holder into the underlying futures contract. (See also *Writer.*)

**Expiration Date*—The time and day on which a particular option may no longer be offset or exercised; *i.e.*, the time it becomes worthless. Also, the month in which options expire is known as the expiration month.

**Hedger*—A professional user of the underlying commodity or a miner of the commodity. A hedger essentially is locking in a futures basis price either by selling against inventory (or mined material) or—in the case of a manufacturer—buying for a future delivery.

Local——A floor trader trading his or her own account only.

**Long*—One who has bought futures contracts or owns the actual commodity.

Margin—The funds put up as security or guarantee for contract fulfillment; collateral. (See also *Variation Margin.*)

**Margin Call*—A request for funds, usually for money, in addition to that deposited as original collateral. (See also *Variation Margin.*)

**Market Order*—An order to buy or sell a commodity at whatever price is obtainable at the time the order reaches the trading ring.

Maturity—The time at which a futures contract can be settled by delivery of the actual commodity.

Member Rate—The commission charged for the execution of an order for a person who is a member of the exchange.

Minimum Price Fluctuation—The smallest price movement possible in trading a given contract.

**Nearby*—The nearest active trading months of a particular commodity futures market.

**Net Position*—The difference between the open commodity contracts long or the open commodity contracts short held by an individual or group.

Nominal Price—A computed price quotation on a future that hasn't traded recently, usually an average of the bid and asked prices.

Notice Day—A day on which notices of intent to deliver pertaining to a specific delivery month may be issued.

Notice of Delivery—A notice provided by the seller giving the details of the delivery he or she will make of the actual commodity. (See also, *Delivery Notice.)*

Offer—An indication of a willingness to sell at a specific price.

Offset—The liquidation or close out of a long or short position by an opposite transaction.

Omnibus Account—An account carried by one futures commission merchant for another futures commission merchant in which the transactions of two or more person are combined rather than separated, and the identities of the individual accounts are not disclosed.

On Track—Commodities that are loaded in railroad cars.

Open Commitment—The obligation assumed by a party to a futures contract.

**Open Interest*—The total number of futures contracts traded in a particular delivery month that have not been liquidated either by an offsetting futures transaction or by an actual delivery.

Original Margin—The initial deposit of margin with respect to a given commitment. (Compare *Variation Margin.)*

Overbought—A technical market situation in which prices are believed to have advanced too far too fast.

Oversold—A technical market situation in which prices are believed to have declined too far too fast.

**Position*—A market commitment. A buyer of a commodity is said to have a *long* position and a seller of a commodity is said to have a *short* position.

**Premium*—The price of an option, the amount of money that the buyer agrees to pay for an option with a particular expiration month and strike price.

**Put Option*—The type of option which gives the purchaser the right to sell the underlying commodity futures contract at a particular price anytime between the purchase date and the expiration of the option.

Rally—An upward movement of prices following an advance.

Reaction—A decline in prices following an advance.

Sample Grade—Usually the lowest quality of a commodity, too low to be acceptable for delivery in satisfaction of futures contracts.

Scale Down—To buy or sell at regular price intervals in a declining market.

Scale Up—To buy or sell at regular price intervals as the market advances.

Scalper—One who trades for small, short-term profits during the course of a trading session, rarely carrying an overnight position.

Settlement Price—The price at which the clearing house clears all transactions at the close of a trading day.

**Short*—A customer becomes short when he or she sells a futures contract with the idea of purchasing it at a lower price at a later date.

**Short-Covering*—A purchase to cover a previous sale.

Speculator—One who buys or sells for his or her own account either as a *day trader* (initiates and offsets his or her positions in one trading day) or a *position trader* (one who holds either a long or short position over a period of days, weeks or months).

**Spot Month*—The futures contract that will come to maturity and become deliverable during the current month.

**Spot Price*—The price at which the physical commodity is selling.

**Spread*—A combination of option positions or option and futures positions for the same commodity in different months. It can be used to offset risk; for example, a long call–short call position would be considered a spread, while a long call–short put position would not, since it does not contain the element of offsetting risk.

Straddle—The simultaneous purchase of one commodity future against the sale of another commodity future.

**Strike Price*—The price at which an option can be converted by exercise into the underlying futures contract.

Tender—Delivery against futures.

Transferable Notice—A notice given by the seller of a futures contract that actual delivery to the buyer will be made. (See also *Delivery Notice.*)

Trend—A general direction of the price movement, either higher or lower.

Variation Margin—The additional funds required if the market has moved against an established position (net). (Compare *Original Margin.*)

Visible Supply—The quantity of a commodity that could be counted.

Volume of Trading—The purchase and sales of futures contracts during a specified period of time.

Warehouse Receipt—A document issued by a warehouse showing possession of the commodity named in the receipt.

Writer (grantor)—The seller of an option. The writer is required to fulfill the option should it be exercised. (See also *Exercise.*)

Bibliography

American Metal Market. *Commodity Trading Supplement*, New York, 1982.

Baruch, Bernard. *My Own Story*, Holt, New York, 1957.

Bernstein, Peter. *A Primer in Money, Banking and Gold*, Vintage Books, New York, 1965.

Browne, Harry. *How You Can Profit from the Coming Devaluation*, Arlington House, New York, 1970.

Bureau of Mines. *Platinum MCP 22, Gold MCP 25, Silver MCP 24, Aluminum MCP 14*, U.S. Department of the Interior, Washington, D.C., 1978.

Casey, Douglas R. *Crisis Investing*, Harper & Row, New York, 1980.

Comex, Inc. *Options on Comex Gold Futures*, New York, 1984.

Commodity Research Bureau. *The Commodity Year Book,* New York, 1980, 1981, 1982, 1983, 1984.

Consolidated Gold Fields, Ltd. *Gold,* London, 1979.

Covelti, Peter. *How to Invest in Gold,* Follett Publishing, Chicago, 1979.

Deutsch, Harry and Effingham, Wilson. *Arbitrage,* 1904.

Friedman, Milton and Jacobson-Schwartz, Anna. *A Monetary History of the United States, 1867–1960,* Princeton University Press, Princeton, N.J., 1963.

Gibson-Jarvis, Robert. *The London Metal Exchange,* Woodless/Faulkner Ltd., London, 1976.

Gold, Gerald. *Modern Commodity Futures Trading,* New York, Revised 1980.

Harlow, Charles V., Jeweles, Richard and Stone, Herbert L. *The Commodity Futures Game,* McGraw-Hill, New York, 1974.

Horn, Fred. *Commodity Futures Investing,* Prentice-Hall, Englewood Cliffs, N.J., Revised 1976.

Jarecki, Henry. *The Golding of America,* Mocatta Metals, New York, 1984.

Kay, Jan. *Investors Advance in Their Pursuit of the Hi-Tech Metals,* New York Mercantile Exchange, New York, 1984.

Kreps, Jr., Clifton H. *Money, Banking and Monetary Policy,* University of North Carolina–The Ronald Press, North Carolina, 1962.

New York Mercantile Exchange. *Metals in Motion,* New York, 1984.

Richenfasher, William F. *Wooden Nickels,* Arlington House, New York, 1966.

Roosa, Robert. *The Dollar and World Liquidity,* Random House, New York, 1967.

Shulman, Morton. *How to Invest Your Money and Profit from Inflation,* Random House, New York, 1979.

Smith, Adam. *Paper Money,* Summit Books, New York, 1981.

Tiffin, Robert. *Gold and the Dollar Crisis,* Yale University Press, New Haven, Conn., 1966.

Index

ABOUT THE AUTHOR

Anthony George Gero is a Vice-President, Investments, of Prudential-Bache Securities, Inc. An active member of the New York Mercantile Exchange since 1966, he is serving his eighth year as a member of the Board of Governors. Currently Treasurer of the Exchange and Chairman of the Finance Committee, he also serves on the Floor, Membership, and Metals (of which he is Vice-Chairman) committees.

Mr. Gero is a member of the Commodity Exchange, Inc., serves on the Floor and Admissions Committee and is a Vice-Chairman of the Control Committee (aluminum) and Broker Training Committee.

A graduate of New York University School of Commerce, Mr. Gero received his Investment Banking Certificate from the Investment Bankers Association at the Wharton School in 1965. He currently teaches a commodities trading course at the New School for Social Research in New York. He is fluent in five languages, and his currency review was published in three of them.